GOLDEN RICE

GOLDEN RICE

THE IMPERILED BIRTH OF A GMO SUPERFOOD

ED REGIS

JOHNS HOPKINS UNIVERSITY PRESS

BALTIMORE

© 2019 Johns Hopkins University Press
All rights reserved. Published 2019
Printed in the United States of America on acid-free paper
9 8 7 6 5 4 3 2 1

Johns Hopkins University Press
2715 North Charles Street
Baltimore, Maryland 21218-4363
www.press.jhu.edu

Library of Congress Cataloging-in-Publication Data

Names: Regis, Edward, 1944– author.
Title: Golden rice : the imperiled birth of a GMO superfood / Ed Regis.
Description: Baltimore : Johns Hopkins University Press, 2019. | Includes
    bibliographical references and index.
Identifiers: LCCN 2019007920 | ISBN 9781421433035 (hardcover : alk. paper) |
    ISBN 1421433036 (hardcover : alk. paper) | ISBN 9781421433042 (electronic)
    | ISBN 1421433044 (electronic)
Subjects: | MESH: Oryza—genetics | Food, Genetically Modified |
    Biofortification | Food Safety | Health Knowledge, Attitudes, Practice |
    Public Opinion
Classification: LCC SB191.R5 | NLM WB 431 | DDC 633.1/71—dc23
LC record available at https://lccn.loc.gov/2019007920

A catalog record for this book is available from the British Library.

*Special discounts are available for bulk purchases of this book. For more information,
please contact Special Sales at 410-516-6936 or specialsales@press.jhu.edu.*

Johns Hopkins University Press uses environmentally friendly book materials,
including recycled text paper that is composed of at least 30 percent post-consumer
waste, whenever possible.

For

Ingo Potrykus

Peter Beyer

Adrian Dubock

CONTENTS

PREFACE

The cover of the July 31, 2000, edition of *Time* magazine pictured a serious-looking bearded man surrounded by a wall of greenery: the stems, leaves, and stalks of rice plants. The caption, in large block lettering, read, "This rice could save a million kids a year."

The man in question was Ingo Potrykus, a professor of plant sciences at the Swiss Federal Institute of Technology, in Zurich, where Einstein had studied and taught. The rice plants around him, although the joint products of many minds and hands, had been largely inspired by him. Their kernels were not the usual plain white grains of rice. Instead, they had a distinct golden hue, the color of daffodils. When spread out on a black surface, they looked like nothing so much as tiny yellow gemstones.

This was Golden Rice, the fruit of nine years of research, experimentation, and development. The "gold" was in fact beta carotene, a substance that is converted into vitamin A in the human body, thus also known as provitamin A. Conventional rice plants already contained beta carotene, but only in their leaves and stems, not in the kernels. Golden Rice also carries the substance in the endosperm, the part of the plant that people eat. This small but portentous change made Golden Rice into a miracle of nutrition: the rice could combat vitamin A deficiency in areas of the world where the condition is endemic and could, thereby, as the magazine headline said, "save a million kids a year."

Vitamin A deficiency is practically unknown in the Western world, where people take multivitamins or get sufficient micronutrients from ordinary foods, fortified cereals, and the like. But it is literally a life-and-death matter for people in developing countries. Lack of vitamin A is responsible for a million deaths annually, most of them children, plus an additional 500,000 cases of blindness. In India, China, Bangladesh, and elsewhere in South and Southeast Asia, many children subsist on a few bowls of rice a day and almost nothing else. For them, a daily supply of Golden Rice could provide enough beta carotene to make vitamin A deficiency a thing of the past. Golden Rice would be a gift of life and sight.

This was a food—indeed, a superfood—that seemed to have everything going for it: It would be the basis for a sea change in public health among the world's poorest peoples. It would be cheap to grow and indefinitely sustainable because low-income farmers could save the seeds from any given harvest and plant them the following season, without purchasing them anew. There is just one thing wrong with Golden Rice: it is a genetically modified organism, and as such is weighed down with all the political, ideological, and emotional baggage that has come to be associated with GMOs—stultifying government overregulation, unreasoning fear, hostility, hatred, dogmatic rejection, and paranoiac horror, much of it based on ignorance of the facts, misinformation, distortions of the truth, and outright lies on the part of environmentalist and other activist organizations and individuals. Greenpeace, for one, was especially vocal in its condemnation of genetically engineered foods, Golden Rice in particular.

As a result of these and other factors, which collectively amounted to an immense retarding force, Golden Rice has not been made available to those for whom it was intended in the 20 years since it was created. To many, this protracted delay is unconscionable, and it brought forth reactions as extreme in their own way as the hyperbolic claims made by GMO opponents. In 2016, for example, George M. Church, a professor of genetics at Harvard Medical School, said in an *Edge* interview:

> Golden Rice was a tough call strategically for Greenpeace and some of their associates. . . . A million lives are at stake every year due to vitamin A deficiency, and Golden Rice was basically ready for use in 2002, so it's been thirteen years that it's been ready. Every year that you delay it, that's another million people dead. That's mass murder on a high scale. In fact, as I understand it there is an effort to bring them to trial at The Hague for crimes against humanity.

Maybe that's justified, maybe it isn't. The fact is we have a pretty good way of addressing this vitamin A deficiency, and nothing else has worked during those intervening thirteen years. It's hard to get them pills because that's very expensive. These are people who could barely afford rice as their sole source of calories; how are they going to afford medicine?

But much of this was sheer overstatement. For one thing, it is doubtful that Golden Rice was "ready," in any but the most technical sense, in 2002. Indeed, some critics would argue that as a proven, viable, agricultural commodity, it is not yet "ready" even today, and, in 2016, one skeptic went so far as to call Golden Rice a "hypothetical product." The fact is that it was invented in 1999 and has been grown, and grown successfully, first in laboratories, then in greenhouses, and finally in open fields, in all the years since. The rice has been subjected to safety studies—toxicity and allergenicity studies—and studies on human consumption, including among American adults and Chinese children, have found it to be more effective in providing vitamin A than spinach and almost as effective as pure beta carotene oil itself. And yet it has not been released for general human use in any country.

Why not?

This book tells the story of how the development, growth, testing, and ultimate release and distribution of this potentially lifesaving biofortified food was delayed by a variety of factors, including sabotage in the form of destruction of test crops in the field by activist groups. But extremist opposition, protests, rhetoric, and even vandalism did not, by themselves, have the power to stop Golden Rice in its tracks or even to substantially hamper the pace of its development.

The first source of delay was simply the scientific and technological difficulty of inventing a new crop type, one that was nutritionally enhanced by molecular methods to express provitamin A in a part of the rice plant that did not normally do so. The tasks of genetically engineering a new metabolic pathway in the plant, getting the plant to express the desired trait at the most beneficial levels of concentration, and then transferring that newly engineered trait into several different varieties of rice successfully—all of these things were, at the time, new, untried, and unproven technologies.

The second cause of delay was the fact that plants themselves are recalcitrant experimental subjects: they grow only so fast and no faster, and the cycle of germination, maturation, and seed production are processes that could not be sped up by conventional means. However, these same processes can easily be slowed down, or even terminated, by a variety of causes such as disease; insect attack; natural disasters and weather events including floods and frosts, heat waves and droughts; vandalism; or simple human misjudgment or mishandling.

But it was something else altogether that had the greatest power to impede the development of Golden Rice, and this was government regulation. That power resided in a complex set of operational guidelines, restrictions, and requirements that constituted an enormous set of obstacles for the Golden Rice scientists to overcome. Governments imposed these constraints in the name of safety, in an attempt to protect human health and the environment. Chiefly responsible for these restrictions is an international treaty known as the Cartagena Protocol for Biosafety and its highly controversial Principle 15, otherwise known as the Precautionary Principle.

This principle states that if a product of modern biotechnology poses a possible risk to human health or the environment, then it is prudent to restrict or prevent the introduction or use of that product or technology, even if the magnitude or nature of the risk is uncertain, speculative, scientifically unproven, or even unknown. Although it may have been benign in its intent, the effect of the principle has been to slow the pace of biotechnology research and development—and in some cases even to halt it, at least temporarily, at multiple times during the research and development process. Those temporary halts and repeated delays had their own pernicious effects, preventing a beneficial product from reaching those who needed it. For them, the wages of precaution were not safety but rather continued illness, blindness, and death.

In the case of Golden Rice, the combined result of these three factors acting together—the scientific difficulty of the project, the slow and stately rate of plant growth and reproduction, and a body of stifling government regulations governing biotechnology research and development—was to prolong the incubation time of a food that, absent externally imposed government restrictions, could otherwise be saving the sight and lives of millions of people.

The story of Golden Rice thus makes for a sad and maddening tale of scientists being repeatedly thwarted in their attempts to invent, improve, breed, field-test, and disseminate a potentially lifesaving and sight-saving food. Despite all these roadblocks, Golden Rice has emerged as the world's first purposefully created biofortified crop. The project began in 1990, when Ingo Potrykus and his colleague Peter Beyer, of the University of Freiburg, started working to genetically engineer a metabolic pathway into a variety of *Oryza sativa*, the world's most commonly consumed rice species, so that the plant's edible kernels would contain beta carotene. It is an understatement to say that their task was daunting. There was no assurance when they started out that what they contemplated was even technologically possible, since it had never been done before. But the two men were highly motivated by the horrors of persistent vitamin A deficiency (VAD) in developing countries, and they viewed their work as a calling, a mission, a quest from which they would not be deterred. Their project was and continues to be a humanitarian one; they were not in it for financial gain or motivated by dreams of selling their miracle plant to industry for a fortune, large or small. Indeed, much of their funding came from foundation grants, from the Rockefeller Foundation, the Bill and Melinda Gates Foundation, and others.

It took almost a decade of laboratory experimentation to invent Golden Rice, but by 1999, Potrykus, Beyer, and a group of colleagues finally succeeded. They inserted a set of genes into the rice genome so that the plant's beta carotene accumulated not only in the plant's leaves and stems, as it normally did, but also in the rice kernels themselves, just as if nature had intended things to work that way from the very beginning.

Once they accomplished that small but powerful technological trick, the inventors naïvely imagined that the hard part was now behind them. Little did they know at the time what was in store for them in the indefinite future. The most difficult tasks, the really hardest stuff, still lay ahead of them. Looking back on it all afterward, Potrykus reflected, "Had I known what this pursuit would entail, perhaps I would not have started."

Once they had their initial proof-of-concept rice in hand, the inventors moved swiftly to develop Golden Rice further, first to improve the product and then to make it available, for free, to poor farmers in developing countries where vitamin A deficiency is a problem of epidemic proportions. In April 2000, they licensed their rice technology to the British agrochemical company Zeneca on a quid pro quo basis: the company retained the right to sell Golden Rice seeds commercially, perhaps as a health food, on the condition that the company financially supported the inventors' future work on the rice and let them distribute the seeds at no cost to small-scale farmers. Zeneca later merged with the Swiss-based company Syngenta, but the terms of the quid pro quo remained unchanged.

On February 9, 2001, Greenpeace, which had a long record of opposition to all GMO foods and crops, issued a statement that an adult would have to eat 9 kilograms (about 20 pounds) of cooked Golden Rice daily to prevent VAD, and that "a breast-feeding woman would have to eat at least 6.3 kilos in dry weight, which converts to nearly 18 kilos [40 pounds] of cooked rice per day." Since the bioavailability of beta carotene in the rice was not then known, there was no factual basis for these claims, which in any case were later proved false. At about the same time, the Indian anti-GMO crusader Vandana Shiva called Golden Rice "a hoax." It was the beginning of a propaganda war against the rice that has only intensified over the years that followed.

But activist opposition was not the real problem. It was the Cartagena Protocol on Biosafety that did the most to impede the optimization, breeding, field testing, release, and use of Golden Rice. The protocol had been adopted in the year 2000 by more than 100 nations, including members of the European Union (but neither the United States nor Canada), for the purpose of protecting human health and biological diversity from potential harms caused by genetically modified life forms and other products of modern biotechnology. The written document, which came into force in 2003, governed the handling, packaging, identification, transfer, and use of "living modified organisms" (LMOs) among the parties to the agreement. One of its provisions, as stated formally in Article 10.6 (and again in Article 11.8), said, "Lack of scientific certainty due to insufficient relevant scientific information and knowledge regarding the extent of the potential adverse effects of an LMO on biodiversity, taking into account risks to human health, shall not prevent a Party of import from taking a decision, as appropriate, with regard to the import of the LMO in question, in order to avoid or minimize such potential adverse effects."

This is one version, among many others, of the Precautionary Principle. Exactly what that principle means in this specific formulation, and how it is to be interpreted and applied in practice, was not immediately clear. It is more of an ideal, a standard of perfection to be aimed at, than a practical, real-world guide to action or public policy. On the one hand, it sounds like a dressed-up variant of a number of innocuous platitudes such as "look before you leap" or "better safe than sorry." On the other, it can equally well be interpreted as a doctrine of "guilty until proven innocent." As one critic put it, the principle "is either so obvious as to be otiose ('if there is cause for concern, be careful'), or so vague as to be meaningless." Nevertheless, the Precautionary Principle became a valuable tool for those who want to stop, or at least delay, any new scientific advance they don't like. In the case of Golden Rice, that delay risked the loss of millions of lives across the years, mostly pregnant women and children.

In light of the Cartagena Protocol, every aspect of Golden Rice development—from lab work to field trials to screening for "regulatory clean events"—was entangled in a Byzantine web of rules, guidelines, requirements, restrictions, and prohibitions. The simple transfer of seed from one country to another became a major logistical problem. "The conditions set up by the Cartagena protocol make exchange of transgenic seed so complicated that it took more than two years to transfer, for example, breeding seed from the Philippines to Vietnam, and one year from USA to India, during which time thirty politically loaded questions were asked in the Indian parliament," said Ingo Potrykus. "These Cartagena conditions are enforced, despite common sense suggesting that it is extremely difficult to construct a hypothetical risk from seed transfer between two breeding stations in different countries, especially for Golden Rice."

For its part, Greenpeace would go on to make a series of often inaccurate negative claims about Golden Rice, for example, "From the outset, the [Golden Rice] project appeared to be designed more towards helping the biotech industry overcome the widespread consumer rejection of GE [genetically engineered] crops, than to help overcome malnutrition" (which was the exact opposite of the truth). Greenpeace claimed that "Golden Rice diverts significant resources away from dealing with the real underlying causes of VAD and malnutrition, which are mainly poverty and lack of access to a more diverse diet" (the implication being that until poverty was wiped out, people would just have to suffer needless blindness and preventable death). And Greenpeace stated that Golden Rice posed "unknown health risks" (but if they were "unknown," then how were they risks?).

Standing against those charges was the fact that Golden Rice was unique among genetically engineered foods, and the properties that made it different also made it immune to many of the conventional criticisms of GMOs. Golden Rice was not invented for profit, and after 2004, when Syngenta renounced all commercial interest in the rice, it would no longer be developed for profit. The rice would benefit the poor and disadvantaged, not modern, multinational corporations. It would be given free of charge to subsistence farmers who can save seeds and plant them from one harvest to the next, without restriction or payment of fees or royalties. The rice was not developed primarily for the benefit of farmers, as were most other GMOs that had been designed to be resistant to herbicides or pesticides. Instead, it was developed for the sole purpose of helping *users*: the malnourished poor suffering from vitamin

A deficiency. And Golden Rice is not a crop upon which a major genetic engineering effort conferred a relatively minor advantage such as a longer shelf life or slightly improved taste, as was true, for example, of the long-since-abandoned Flavr Savr tomato.

Rather, Golden Rice is a food that can actually *save lives and prevent blindness*, and it can do so safely, economically, and sustainably. "The great advantage of GMO seeds is the fact that the entire technology is embedded in the seed," Potrykus has said. "All a farmer needs to benefit from this technology is one seed. . . . Each seed can produce 20,000 metric tons of Golden Rice within two years."

Had Golden Rice been left free of the restrictive regulatory conditions, requirements, and political opposition that have hampered its development, cultivation by rice farmers, and distribution to users in some of the poorest regions of South and Southeast Asia, millions of lives would not have been lost to malnutrition, and millions of children would not have gone blind.

Vitamin A deficiency is one of the world's biggest killers of children, and it's caused merely by the lack of an essential nutrient that is required by the body in only small quantities. If the deaths of these children is not "a crime against humanity," it is nevertheless a modern tragedy.

GOLDEN RICE

CHILD KILLER

For people in the rich, well-developed countries of the Western world, vitamin A deficiency could scarcely be called a clear and present danger. There, it's not deficiency but rather wretched excess that is the major health problem. That fact is all too manifest in the current obesity epidemic, which is an artifact of people having access to, and eating, too much food and, often enough, foods of the wrong type. And so for all intents and purposes the concept of a micronutritional deficit just doesn't exist for Westerners. To get a realistic sense of what such deficiencies mean in practice, we must look elsewhere, to lands remote and foreign to the Westerner. In her book *Vitamania*, Catherine Price takes us to exactly the right place, and at the right time of day.

> Visit a village in many areas of sub-Saharan Africa or South Asia at twilight, and you may notice something strange: groups of children who have played together all afternoon will bifurcate at dusk. Some will continue their games, running around in the fading light, while others will retreat to their families' huts, sitting with their back to the corner. They won't reach for nearby toys; they won't even move for food. Instead they will remain in place, eyes blankly staring, until a friend or family member brings food to them or guides them away by the hand.
>
> The reason for their stillness is simple: they can't see.

Their condition is called night blindness (nyclatopia), and it's one of the first indications of vitamin A deficiency. Night blindness is by no means confined to children, however: local women in the later stages of pregnancy often suffer from the condition. As the light dims, their visual field gradually fades to black. This is not pleasant.

Among the many other roles it plays in the body—it is known to influence the expression of more than 300 genes—vitamin A is necessary for night vision. Another name for vitamin A is retinol, so called for its crucial function in the retina. The surface of the retina is made up of a complex array of light-sensitive photoreceptor cells: rods and cones. The cones are responsible for daylight vision and color perception. The rods operate in dim light and give us an essentially monochromatic view of the world. The rods contain the pigment rhodopsin, also known as visual purple. In normal functioning, light photons striking the rods break down the rhodopsin into its two chemical components, opsin (a protein) and retinal, one of three active forms of vitamin A.* That chemical reaction gives rise to signals that are transmitted by the optic nerve to the brain, which converts them into the visual images that we see.

The rhodopsin that's broken down in this fashion has to be recycled, with its two chemical components, opsin and retinal, being rejoined in order to function again. But a portion of the retinal is lost in the separation process and must be resupplied from outside the body or from the body's own stores of vitamin A in the liver. The human liver can hold large amounts of retinol—enough to last a year or more for those eating a healthy, vitamin-rich diet. But those who are vitamin A deficient have no such reserves to call upon. Their rhodopsin cannot be rebuilt, and their night vision suffers accordingly. If the deficiency persists, the rods lining the retina deteriorate and cease to function, and tiny white lesions appear on the surface of the retina. The night blindness will last for as long as the deficiency persists, but the condition is easily reversible by restoring the vitamin to normal levels.

Nutritional blindness is one of the oldest illnesses known to medicine. An ancient Egyptian medical papyrus—the Ebers Papyrus, which dates back to 1550 BCE—contains both a description of the phenomenon as well as a prescription for an effective cure: the eating of animal liver, which

---

* Vitamin A consists of a family of retinoids. The three active types are retinol (the alcohol form), retinal (the aldehyde form), and retinoic acid (the acid form).

is rich in vitamin A. Ancient Babylonians and Greeks also prescribed this remedy, although why the practice worked remained unknown until the early twentieth century, when the vitamin was first discovered.

Counterintuitively, while night blindness is one of the first *symptoms* of vitamin A deficiency, it is actually a late *stage* of a much more serious set of conditions. Vitamin A plays a large multiplicity of roles in the human body: it's necessary to maintain the skin, as well as the epithelial cells that line many body cavities and tubular organs including the intestines, lungs, and respiratory and urinary passages. It helps maintain the immune system and our ability to fight infections; it promotes bone growth and helps preserve the teeth; it assists in reproduction and is important for embryonic development. Lack of the vitamin also worsens the effects of common illnesses such as diarrhea and childhood diseases such as measles. It's quite remarkable, the range and variety of functions of this substance that, if absent from the body, produces so many serious, and even lethal, ill effects.

Because they occur internally and out of sight, many of the effects of vitamin A deficiency are insidious. One of the first to become acquainted with this fact was a Johns Hopkins ophthalmologist, Alfred Sommer, who in the early 1980s was doing a study of preschool-age children in rural Indonesia. He examined the eyes of some 3,000 children every three months for a year and a half and was surprised to find that children with either mild night blindness or Bitot's spots (pearly white, foamy buildups on the surface of the eye) had 4 times the average mortality rate, and in some age groups up to 8 to 12 times, of children without those conditions. This suggested to him that vitamin A deficiency was much more than just an eye problem. Apparently, it could kill you.

If left untreated, night blindness is followed by a condition known as dry eye (xerophthalmia), in which the surface of the eye and the cornea become inflamed and scaly. It progresses to perforation of the cornea, an open wound on the eye that can let the inner gel, the vitreous humor, leak out, leading in many cases to total and permanent blindness, and finally to loss of the eye. Meanwhile, many of the more pernicious effects of vitamin A deficiency are occurring inside the body, which is why the eye problems of the children examined by Alfred Sommer all too often ended in death. Vitamin A deficiency is the leading cause of preventable blindness in children, and about half of affected children die within a year of becoming blind.

■ Vitamin A deficiency is one of the world's biggest killers of children. In 2010, according to one estimate, this specific deficiency killed more children than HIV/AIDS, tuberculosis, or malaria—about 700,000 boys and girls under age five per year, almost 2,000 deaths per day. On a global basis, about one-third of children under five suffer from the condition.

Reliable and accurate estimates of the worldwide incidence or death rates from vitamin A deficiency are hard to come by, however. Numbers vary over the course of time and come from a variety of sources including the World Health Organization, UNICEF, the National Institutes of Health, and other institutions, which in turn rely on data supplied by national governments. Since some countries do not even collect national disease prevalence information, it follows that any given reported figure is almost certainly an underestimate.

In 1987 the World Health Organization reported that vitamin A deficiency was endemic in 39 countries. This estimate was based on eye examinations and on tests for retinol concentrations in the bloodstream. In 1995, WHO updated this figure and stated that the deficiency was a problem of "public health significance" in 60 countries and was "likely to be a problem" in 13 additional countries. These higher numbers reflected an increase in submitted data, that is to say, better reporting, rather than an increase in the actual incidence of the disease.

In 2005 WHO again revised its figures, stating that vitamin A deficiency was a problem in a total of 122 countries. Further, it estimated that night blindness affected 5.2 million preschool-age children and 9.8 million pregnant women. Tests of blood serum levels, however, revealed an even worse picture: "Low serum retinol concentration affects an estimated 190 million preschool-age children and 19.1 million pregnant women. This corresponds to 33.3% of the preschool-age population and

15.3% of pregnant women at risk of VAD, globally." Those most affected were in Africa and Southeast Asia.

A separate estimate, based on a 2010 UNICEF report, stated that providing adequate vitamin A to children in undernourished settings would prevent from 1.9 million to 2.7 million child deaths annually. By any standard, that is a lot of preventable deaths.

Vitamin A deficiency is not a disease in the usual sense of an illness caused by the presence of a foreign agent such as a bacterium, virus, parasite, or a toxin. The illness is not transmissible from person to person. It is not a genetic disease. Rather, it is a disorder caused simply by the lack of an essential substance, retinol, which is needed in only very small amounts to prevent illness. The US recommended dietary allowance of retinol for children aged nine to thirteen is 600 micrograms per day. That is an amount approximately equivalent by weight to one grain of table sugar.

How can the absence of something needed in such incredibly small quantities cause so many diverse illnesses, including blindness and death? For that matter, how was it ever discovered that any illness can be caused not by the *presence* of something, like a virus, but rather by the *absence* of something, like a biochemical?

And how did scientists ultimately find the missing entity? How do you discover something that isn't there? Nonexistence, after all, leaves no traces.

But it does have effects.

It's this hidden, shadowy quality of vitamins that explains why, as a category of substances that are essential to health, micronutrients were not discovered until the early years of twentieth century. It also explains why, according to vitamin scholar and historian Richard Semba, the discovery of vitamins was not "a mythical story of crowning scientific breakthroughs, [but rather] was a slow, stepwise progress that included setbacks, contradictions, refutations, and some chicanery." There was no dramatic "eureka" moment, no solitary genius or towering medical hero who could lay claim to having found the promised land. When it finally did occur, however, the discovery of vitamins constituted a major advance in the scientific understanding of health and disease.

The term itself, originally "vital amine" and "vitamine" (from the Latin *vita* for life + *amine*, since vitamins were thought to contain an amino acid), was coined by the Polish biochemist Casimir Funk in 1911, even before the first vitamins had been purified and isolated. But the slow and halting process of bringing to light these invisible substances that lurked in foods started long before that.

One of the first great leaps of intuition pertained to the well-known shipboard disease, scurvy. Scurvy also occurred in towns under siege as well as in other circumstances where food was in short supply or was nutritionally inadequate. In 1747 the British naval surgeon James Lind made the mental connection between the disease and diet. Moreover, he had a way of demonstrating the linkage, for while serving aboard the ship HMS *Salisbury* he had a captive audience, making it an ideal setting for a controlled experiment (regarded by some as the first-ever clinical trial). Lind divided up a group of 12 sailors suffering from scurvy into six pairs and fed each pair a diet that included one of six purported scurvy cures. It turned out that the two crew members who ate oranges and a lemon every day had the quickest and most complete recovery. "Oranges and lemons," Lind concluded, "were the most effective remedies for this distemper."

But why? There were several possible explanations. Perhaps there was a noxious agent in the body that caused scurvy, and the acids in the fruit destroyed it. Possibly there was some putrefaction or decay in the patients, and something in the fruits removed it. It was only logical to think that some harmful presence caused the disease and that an unidentified substance in lemons and oranges destroyed it. In the end, Lind decided that both the cause of the disease and the mechanism of its relief were unknown and "merely conjectural." And, indeed, almost 200 years would elapse before the identity of the curative agent, vitamin C (ascorbic acid), was determined.[*]

_____

[*] Vitamin C was discovered by two biochemists, independently and more or less simultaneously, in the 1930s: the American Charles Glen King and the Hungarian Albert Szent-Györgyi.

Although the story is less well known, vitamin A deficiency, as manifested in the form of night blindness, also occurred during the course of prolonged ocean voyages. The symptoms appeared gradually, with sailors being unable to see objects shortly after sunset or in clear moonlight. And then even in the brightest candlelight the affected seamen would stumble around like blind men, grope their way past objects, and bump into each other.

Naval surgeons thought they knew exactly what the problem was: overexposure to light. After all, the sailors stood watch in bright sunlight, with the open sea, the ship's many white sails, white clothing, the deck, and metal fittings all reflecting light rays back at them, the combined effect of which "dazzled the eyes." The obvious solution was to confine affected sailors to dark quarters below decks—sometimes for weeks at a time. The seamen of course hated this treatment, which anyway never really worked. It was puzzling that the condition affected only common seamen, never the ship's officers, some of whom kept livestock aboard to provide themselves (but not the ordinary sailor) with fresh eggs, milk, and cheese. Therefore, in 1860 the British physician Alexander Bryson theorized that night blindness had nothing to do with overdazzled eyes but was "entirely dependent on an improper or erroneous diet."

Proof of a dietary cause of night blindness, as well as the increased mortality rates associated with it, came not from shipboard experiments, however, but from observations made in clinical settings on dry land. The process began in 1816 when the French physician François Magendie fed dogs a regimen of nothing but water, sugar, gum arabic, and little else, in particular, no proteins. The dogs wasted away, developed the corneal ulcers characteristic of vitamin A deficiency, and died. In 1828 another French physician, Charles-Michel Billard, noted that the corneal perforations observed by Magendie were remarkably similar to those that Billard himself found in poorly fed infants in the Hôpital des Enfants-Trouvés in Paris. What these cases had in common was that the abnormal eye conditions were correlated with nutritional deficiencies: something was absent from the diet, but just what was still unknown.

The discovery of vitamins then got sidetracked by one of the most successful constructs in the history of medicine, the germ theory of disease. Originating with Louis Pasteur in 1865, it claims that diseases are caused by tiny organisms (or their toxins), which in some cases can be spread from person to person. This was a sweeping, powerful theory, and evidence in its favor continued to mount as researchers connected diseases to their microbial causes, as observed in microscopes. For example, Pasteur himself discovered the parasite that caused silkworm blight (the culprit was a little-known fungus). Other scientists identified the microbial causes of diseases including typhoid fever, tetanus, tuberculosis, cholera, malaria, bubonic plague, anthrax, and diphtheria.

But, as successful as it was in pointing to the causes of a large variety of diseases, the germ theory was not truly all encompassing, and a small group of illnesses continued to elude explanation by pathogenic agents. Scurvy was one; others included beriberi, rickets, and pellagra, as well as night blindness, corneal ulcers, and dry eye syndrome.

Deficiency diseases are products of deprivation, and in 1870 the Siege of Paris provided a fresh opportunity for scientists to study the effects on the human body of the lack of a particular foodstuff, in this case, milk. The Prussians were so successful in isolating the city from its normal food supply that residents were soon reduced to eating horsemeat, cats and dogs, rats, and finally even zoo animals, including camels, zebras, kangaroos, and two elephants named Castor and Pollux. The city's entire population suffered, and the death rate increased from 900 per week to more than 4,000. But it was the infants and young children of Paris, as well as breastfeeding mothers, who were most at risk from the lack of cow's milk from farms in the surrounding countryside.

In response to this situation, the chemist Jean-Baptiste-André Dumas synthesized a crude milk substitute, but infants fed this artificial milk continued to waste away and die of starvation. Evidently there was something missing from his manufactured milk, some element that, as Dumas wrote, "the smallest and most insignificant traces of [which] may prove to be not only efficacious, but even indispensable" to life.

It was the German physician Wilhelm Stepp who zeroed in on the substance in question. In the early 1900s, he performed an ingenious series of experiments in which he fed mice a nutritionally adequate diet, including milk, and then selectively removed certain ingredients one by one. The mice began to lose weight when a fat-soluble substance of un-known composition was removed from their food, but they recovered and regained weight when the mystery substance was resupplied. An-other researcher, Frederick Gowland Hopkins, claimed in 1912 that these fat-soluble substances in milk were "accessory factors" necessary "in as-tonishingly small amounts." This was a close approach to vitamin A, which is indeed a fat-soluble compound.

And then, in an endgame race that proceeded so slowly and circu-itously as to be almost painful, a small cadre of scientists, some in open competition with the others, some working cooperatively, and one mak-ing a fraudulent priority claim, each inched their way toward "discover-ing" vitamin A. In 1913, two scientists at the University of Wisconsin, Elmer McCollum and Marguerite Davis, claimed, based on experiments with rats, that an "extract of butter or of egg" was necessary to normal growth. They were "convinced that these extracts contain some organic complex without which the animals cannot make further increase in body weight." But they nevertheless could not say what that organic complex was and in fact concluded that "the further study of the nature of the 'active' bodies in these extracts must of necessity require a great deal of time and labor." Despite this admission, McCollum would later claim to have discovered vitamin A. He did not.

In 1916, McCollum proposed calling this still "unknown substance . . . fat-soluble A." Four years later, in 1920, the British biochemist Jack Drum-mond called it instead "vitamin A." We now had a substance and a name, but no chemical structure that answered to it and no pure sample of it in hand.

Finally and at long last, in 1931 the Swiss chemist Paul Karrer isolated and established the chemical structure of vitamin A, a feat that won him the 1937 Nobel Prize in Chemistry. In awarding it, the Nobel Committee stated that "Karrer succeeded in extracting vitamin A from cod-liver oil and in determining its composition." And so if any one person could be said to have "discovered" vitamin A, it would be he. More realistically, though, the vitamin was discovered not by any one person acting alone but rather by the collective body of researchers, physicians, and chemists whose experimental work and scattered observations spanned the 115-year period between Magendie's feeding trials on dogs in 1816 and Karrer's analysis of retinol's chemical nature in 1931—a long and tortuous chain of events.

■ Once retinol had been isolated, its structure determined, and its role in preventing micronutrient deficiency diseases had been established, it would seem that the solution to the problem of vitamin A deficiency was self-evident: just supply the vitamin to people who needed it. There were three main forms of providing it: supplements (capsules, syrups, or oils), fortified foods, or vegetables high in vitamin A.

But all these methods, no matter how straightforward they might appear at first glance, had their own respective drawbacks. Supplementation, for example, had a number of problems. To begin with, for all of the benefits it provides to human health when administered properly, vitamin A is toxic in high doses. When taken during pregnancy, large doses of the vitamin can cause fetal and congenital birth defects, including malformations of the eye, skull, lungs, and heart. In children, overdose can cause transient nausea, vomiting, headache, and dizziness—side effects that disappear within a day or two and develop in only about 5 percent of cases. Young infants (who are born with stores of vitamin A sufficient for only the first few days of life) may also suffer from bulging fontanelles (the spaces between skull bones), a condition that usually resolves by itself within 72 hours. And, in rare instances, high doses of the vitamin can even cause death. Still, these risks are relatively small, especially as compared to the far greater risk of dying of vitamin A deficiency itself, and are approximately on par with the chance of adverse reactions to childhood vaccinations for diseases such as diphtheria, pertussis, and polio.

Second, there was the concern among some health professionals and nutritionists about whether supplementation with synthetic vitamin A was "natural." The International Vitamin A Consultative Group (IVACG), for example, was established in 1975 with the goal of fighting vitamin A deficiency disorders, but the idea of using supplements was not popular with some of its members. Alfred Sommer, the ophthalmologist who in the 1980s discovered in Indonesia that night blindness and dry eye were signs of much more severe health problems, recalled that "discussions at IVACG remained acrimonious for many years, riven with fruitless arguments over what constituted a *natural* solution to the problem. . . . Supplementation was opposed by many nutritionists as an 'unnatural' act, insisting, unhelpfully, that the real answer was to get people to grow and consume more green leafy vegetables."

Nevertheless, at length that obstacle was overcome, and a vitamin A supplementation program with aspirations for global coverage began in 1993, under the auspices of UNICEF and other aid organizations. (Sporadic programs had begun even earlier, sponsored by the governments of countries including Indonesia, the Philippines, and Vietnam.)

But even when supplements were made available to children on a mass scale, vitamin A deficiency persisted. According to a UNICEF estimate, during the eight-year period from 1997 to 2005, more than 4 billion vitamin A capsules were distributed, a course of action that saved the lives of at least 3.5 million children. Despite the delivery of those 4 billion capsules, though, 2005 was the year in which the World Health Organization estimated that vitamin A deficiency remained a public health problem in 122 countries and affected an estimated 190 million preschool-age children as well as 19.1 million pregnant women.

Supplementation programs, in other words, had their flaws, one of which was the problem of ensuring adequate, much less universal, coverage. It was difficult if not impossible to get the capsules to all who needed them, particularly those in poor, outlying, rural areas, the very children and expectant mothers who are often the most at risk of serious and prolonged deficiencies. Further, the need for vitamin A is lifelong, meaning that there had to be no interruption of a twice-yearly administration of the proper dose, sustained across an entire lifetime. Also, supplement programs required a large investment in human resources to keep track of those receiving the doses, to manage vitamin A supplies, and to monitor the effects on child health and survival.

Writing in 1995, four years before the advent of Golden Rice, Alfred Sommer and Keith West summarized the problems with supplementation: "Difficulties most often encountered include recurrent depletion of central stores, transport delays and breakdowns in local procurement that lead to peripheral capsule shortages, poor supervision and motivation of local workers[,] . . . and inadequate local coordination and record-keeping. Children may be incorrectly targeted (i.e., outside the age group) or receive the wrong dose."

The second possible, and oft-proposed, solution to vitamin A deficiency was "to get people to grow and consume more green leafy vegetables." This was a "natural" solution, but it, too, had its limitations. First of all, since many of those who are vitamin A deficient are also landless, they have no easy way of growing their own vegetables. But, even where they did, the grow-your-own approach did not invariably lead to better nutrition. One study of an 11-year-long home garden project in Senegal in fact showed a surprising *negative* correlation between growing vegetables and nutritional status. "This negative nutritional effect was attributed to the selling of produce and the use of profits for non-food purposes."

For those who don't have access to a home garden, the only way to get the needed produce is to buy it. But these relatively expensive produce items, day after day, in the quantities needed to prevent deficiency, are out of reach for these severely impoverished populations. Furthermore, the bioavailability—the proportion of a nutrient contained in food that is actually absorbed by the body when eaten—of vitamin A in many fruits and vegetables is not high. According to bioavailability studies of plant foods like spinach, a child between the ages of one and three would have to eat eight daily servings of dark green leafy vegetables to meet the recommended dietary allowances for vitamin A.

Solving the problem by means of animal-sourced foods such as eggs, cheese, and butter is also not a realistic solution for target population groups. Most vitamin-deficient people are so poor and disadvantaged that foods like butter and eggs are luxuries, and not remotely everyday fare.

■ The remaining option, fortification, is the process of adding micronutrients to foods over and above the levels that they might already contain. Iodized salt, vitamin D fortified milk, and orange juice with added calcium are examples of fortified foods. In many ways, this is the best, most elegant, and most practical solution to the problem of vitamin A deficiency. Fortified foods are nontoxic, meaning that you can't overdose on them. If the target population already consumes the food item on a regular basis, no vast infrastructure is required to distribute and administer supplements twice yearly for the lifetimes of all the millions of recipients.

But fortification programs have had a mixed record of success. Several foodstuffs have been fortified with vitamin A, including sugar, wheat, dairy foods, MSG, and margarine. MSG fortification was used for a while in the Philippines, but the cost of the fortification was passed on to consumers, which prompted them to reject the fortified MSG packets in favor of larger, nonfortified versions. There were also some problems with the chemical stability of the MSG+A over time.

Some countries in Latin America started to fortify sugar, with positive health outcomes, but continuance of the programs fluctuated because of civil conflict, lack of enforcement of the fortification law, variations in the cost of sugar, and other factors.

In their book, *Vitamin A Deficiency*, Alfred Sommer and Keith West concluded, "Implementing a national fortification program is a major undertaking that requires sound scientific rationale, industrial capacity, training advocacy, adequate legislative support, economic viability, community acceptance and long-term sustainability, monitoring and quality control."

For the world's poor, the ideal food to fortify is their default dietary staple, rice. Rice is eaten daily by about half the world's population. It's available widely and cheaply. And so if there were ever a food tailor made to combat vitamin A deficiency, it is *biofortified* rice, rice that already contains the added micronutrient as it grows. It would not have to be added manually, as iodine is added to salt during processing or vitamin A is added to MSG. It would require no record keeping or expenses, over and above those that already pertained to ordinary rice.

This is exactly what Golden Rice is: a biofortified rice whose kernels contain enough beta carotene to prevent vitamin A deficiency by the simple means of eating the amount normally consumed on a daily basis. The product would be cheap, easily available in adequate quantities, and, since seeds could be saved and replanted repeatedly, self-sustainable. After all, seeds reproduce; pills do not.

If there were ever a magical, easy, and cheap way to help alleviate the problem of vitamin A deficiency, a solution that would go of itself, that solution is Golden Rice.

2
PROOF OF CONCEPT

The original idea for Golden Rice didn't come from an industrial or academic biotech lab, whether in the United States, Europe, Asia, or anywhere else. Nor did it emerge from a food, seed, or farming company. The idea first arose from within a traditional rice breeding organization in the Philippines, the International Rice Research Institute (IRRI, pronounced "eerie"), located in the city of Los Baños, about 40 miles south of Manila. IRRI had been founded in 1960 by the Ford and Rockefeller Foundations together with support from the Philippine government, as a center for study of the rice plant, "with a view to increasing production per unit area, both in quality and quantity."

Rice is one of the world's most important foods, a staple for almost half of its peoples, many of them in developing countries. The great profusion of rice varieties reflects its global popularity, as residents of different countries each cultivate and eat their own locally preferred types. While the total number of rice varieties is not known with precision, it is estimated that there are at least 40,000 types of the single most cultivated rice species, *Oryza sativa*. There is long-grained, short-grained, and round-grained rice; sticky rice and nonsticky rice; dwarf, semidwarf, and tall rice; fragrant rice and nonaromatic rice. In addition, different rice cultivars have been selectively bred for the wide range of conditions under which they are grown. For example, some types are more drought tolerant or disease or insect resistant than others. There are differences in yield, grain quality, taste, and other properties, and breeders are always looking for ways to improve the rice plant and to modify its kernels.

When IRRI first began operation in the sixties, there was one principal method for producing new types of rice, crossbreeding one variety with another. Developing an improved type of rice was not a new, un-usual, or even particularly challenging process. The object of traditional rice breeding, after all, was to create novel variants that were better than the old ones on some scale of value, and this had already been done thousands of times by the world's rice farmers, conceivably stretching back to the very beginnings of rice agriculture. Genetic evidence has established that all the different forms of Asian rice emerged from a sin-gle domestication event that occurred about 10,000 years ago in the Pearl River valley region of China. More recently, breeders working at IRRI have themselves developed 843 new rice varieties across the 60 years of the institute's existence and have released them to 77 countries.

In general, creating a new plant type is a relatively straightforward business. When one or more individual plants out of a given crop exhib-its a striking variation—better looking, better tasting, more productive in bearing fruit—then any tiller of the soil, farmer, planter, or home gar-dener can preferentially save the seeds from those plants and sow them the next season. Those variant plants—called "sports"—are often prod-ucts of spontaneous mutations, and their selective cultivation brings a new breed of plant into existence.

Another common breeding technique is cross-pollination, transfer-ring the pollen from one plant to one or more others, a process so basic and "natural" that such transference occurs without human interven-tion, by bees, birds, bats, or the wind. A farmer does it manually, trans-ferring pollen grains from plant to plant, creating a new strain by dupli-cating what unaided nature has already done millions of times before. Hybridization is a special type of cross-breeding between two widely dissimilar parent plant types, sometimes even between two different species. All of this is ancient history.

But about a dozen years after IRRI was founded, a radically new breed-ing technology was just getting under way. This was genetic engineering, which allowed scientists to make changes in a plant's (or an animal's) ge-nome directly, by the use of molecular tools.

The genetic engineering of living organisms is generally understood to have begun in 1973, when the biochemists Herbert Boyer (of the University of California, San Francisco) and Stanley Cohen (of Stanford) changed the genetic makeup of a bacterium, *E. coli*, to be resistant to the antibiotic kanamycin. Key to the process was a cellular structure known as a plasmid.

Plasmids are small circular stretches of DNA that float around freely in the cytoplasm of bacteria and replicate independently of the chromosomes. These cellular structures would become workhorses of genetic engineering. In the end (though not in the beginning) they were used by Ingo Potrykus and Peter Beyer in the invention of Golden Rice.

What makes plasmids so useful in genetic engineering is the fact that you can open up these tiny circular structures, insert a new gene into the gap, and then close the circle back up again. The opening of the plasmid is made possible by a particular class of enzymes called restriction endonucleases, while the reclosing is accomplished by another enzyme, a ligase, which is in effect a biomolecular glue.

In 1973, Cohen and Boyer cut open a range of plasmids in this way, inserted a kanamycin resistance gene into the gap, and then closed up, or recombined, the two open ends. The reconstructed plasmids now contained the new genetic sequences. Finally, they put the new and improved plasmids back into the *E. coli* bacteria from which they came and grew them out on culture plates.

If everything worked as planned, then any of the cultured *E. coli* without the plasmid containing the kanamycin resistance gene would be killed off when exposed to the antibiotic, but the bacteria that had successfully taken up the resistance gene would be unaffected. So long as the antibiotic resistance gene had been properly inserted into the plasmid and the protein it coded for was correctly expressed, the bacterium would be immune to kanamycin. Although the underlying process was different, the effect was akin to being vaccinated against a disease.

And that is exactly what happened to the transgenic *E. coli*: they had become resistant to kanamycin. This was a milestone experiment in molecular genetics, and it was noteworthy for at least two reasons. One, as artificial and "unnatural" as it might appear, the act of putting new genes into organisms is a process that bacteria already engage in, naturally and normally, all by themselves. As the late evolutionary biologist Lynn Margulis once wrote (if a bit hyperbolically): "Bacteria trade genes more frantically than a pit full of commodities traders on the floor of the Chicago Mercantile Exchange." They do this in two ways, one of which is a recognizable version of bacterial sex, when one bacterium sends out a tubular protein bridge to another bacterial cell and through it donates some of its genetic material to the recipient. This is called *conjugation*, and there is even a lurid three-way version in which three bacterial cells are connected simultaneously.

But there is a second version of bacterial gene exchange, and it more closely approximates the plasmid insertion technique that was used by Cohen and Boyer. In this second method, a bacteriophage virus attaches itself to the outer surface of a bacterium and injects a foreign gene from another organism. This process is called *transduction*. A difference between the two methods is that transduction does not require physical contact between the cell donating the DNA and the cell receiving it, as in conjugation. Rather, the phage virus takes the foreign gene from one bacterium, shuttles it away, and then inserts it into another one. A phage virus that does this is called a *vector*.

Taken together, conjugation and transduction are ubiquitous in nature; they are methods of rapid gene transfer and genetic recombination that occur regularly and automatically, all the time. And, just as the process of manual cross-pollination duplicates something that nature already does, Cohen and Boyer's method of transferring genes between organisms essentially duplicates the bacterial process of transduction.

Their genetic engineering feat was noteworthy also in that it constituted the basis of a new kind of manufacturing system. *E. coli* bacteria reproduce themselves every 20 minutes or so, which means that a colony of such organisms could be used to make large quantities of useful substances, as Cohen and Boyer realized. "Genetic manipulation," Cohen said, "opens the prospect of constructing bacterial cells which can be grown easily and inexpensively, [and] that will synthesize a variety of biologically produced substances such as antibiotics and hormones."

In 1976, Boyer founded the firm Genentech to put this technology into commercial practice. Six years later, in 1982, the company received FDA approval for Humulin, human insulin produced by bacteria. The product was made by the method that Cohen and Boyer had pioneered. The genetic code for human insulin was inserted into plasmids, which were then placed into bacteria, which synthesized the drug as they reproduced themselves.

Despite being a product of genetically engineered bacteria, the resulting insulin is nevertheless genuinely *human* insulin, as opposed to the insulin in use by diabetics until then, which was extracted from the pancreatic glands of pigs and cattle. Humulin has a number of advantages over animal-derived insulin: It cannot cause insulin allergic reactions, as are sometimes caused by insulin from animal species. It is purer than pig or cattle insulin. It has a slower onset of action as well as a longer duration of activity. And although it was initially more expensive than animal insulin, it soon became much cheaper.

For all these reasons, genetically engineered human insulin became widely accepted in the treatment of diabetes. There were no mass protests or demonstrations against this genetically engineered "Frankendrug," no hostile press releases by Greenpeace, no efforts to stop production or to dump batches of the drug into open fields, pits, or rivers, or to destroy it in huge bonfires. Indeed, this was one genetically engineered product that seemed to provoke no opposition.

Golden Rice is also a product of genetic engineering and a substance that enters the human body, where it can preserve and enhance health and save sight and lives. But, unlike Humulin, Golden Rice would enjoy no such immunity to criticism.

■ But first it would have to be invented.

In April 1984, a group of rice scientists gathered at IRRI for a five-day seminar to learn about the potential applications of genetic engineering techniques in agriculture, specifically to the improvement of rice. The seminar, which was called "Biotechnology in International Agricultural

Research," had been organized by Gary Toenniessen, who was then assistant director for agricultural sciences at the Rockefeller Foundation in New York. The year 1984 marked the start of the foundation's $100 million, 15-year-long biotech rice research program.

At that point, molecular biology had been around for about a decade and had even been applied to some agricultural products, but not yet to rice. However, by means of the direct importation of novel genetic sequences into the rice genome, it would now be possible, at least theoretically, to create rice plants with qualities impossible to achieve by means of conventional breeding. Many of the rice breeders attending the seminar were therefore looking forward to talking about, thinking about, and manipulating genes, the tiny, invisible entities that gave any rice variety its phenotype, its set of observable characteristics.

At the conclusion of one day's conference presentations, Gary Toenniessen and a group of rice breeders were sitting around at the IRRI guesthouse drinking beer. As it happened, a few of those present were somewhat doubtful that these new molecular techniques being discussed could have a significant or lasting impact in producing better types of rice. So Toenniessen asked the group: Suppose that it *was* actually possible to insert into the rice genome a novel gene that coded for a new and desirable characteristic. In that case, what new gene would you choose? "What gene would you put into rice, if you could put in any gene at all?"

Most of the answers were fairly predictable: genes for resistance to rice blast disease, drought tolerance, insect resistance—basically just enhancements of qualities that were already present in dozens of rice types. But finally it was Peter Jennings's turn to answer the question.

Jennings was, at that time, IRRI's star rice breeder. He had worked there since 1961 and had produced the first "miracle rice," also known as "the rice that changed the world." This was a rice variety called IR8. In 1962 Jennings had made 38 crosses of different varieties from types already in the inventory at IRRI. The eighth cross was between a Chinese dwarf variety called Dee-geo-woo-gen and a tall Indonesian variety called Peta. This produced a good-looking plant but one that had yielded only 130 individual seeds.

But Jennings kept replanting generation after generation of this eighth cross until, in the fourth filial generation, he got a plant type that matured in 130 days, as opposed to the more usual 170 days. Further plantings revealed that this new rice type also produced five tons of rice per hectare, which was 10 times the traditional rice yield. Thus, the "miracle rice," named IR8, indeed had ripple effects in increasing rice production in many countries around the world.

So when it came time for Jennings to say what gene he would add, everybody listened.

"A gene for yellow endosperm." His answer was pretty baffling until he explained why. "As long as I've been a rice breeder—over twenty years—I've been looking for a rice with yellow endosperm," he said. "Because then it would produce beta carotene, the precursor of vitamin A. When children are weaned, they're often weaned on rice gruel. And if they don't get any beta-carotene or vitamin A during that period, they can be harmed for the rest of their lives."

That, then, was the birth of the idea for yellow rice. And it now made sense to Gary Toenniessen, who held a PhD in microbiology and knew that other plant species *do* express beta carotene in their endosperm: maize, for example, whose kernels are yellow because they contain beta carotene. Beta carotene is a plant pigment that comes in various yellow, orange, or red hues. But Toenniessen also knew that maize and rice belonged to different genera and therefore did not cross-hybridize; that is, you can't put maize beta carotene genes into rice grains with conventional cross-breeding methods. But you could conceivably do it by using recombinant, molecular techniques. At a minimum, you'd need to identify whatever gene (or genes) in maize, or some other plant, coded for beta carotene, and then you'd have to clone that gene, transfer it into the rice genome, and then activate it.

Easy enough to say, and conceptually simple, but actually doing it in practice was quite another thing. But Toenniessen was in charge of a veritable funding machine at the Rockefeller Foundation, and so he launched a research project into how to make the necessary molecular alterations. First he commissioned a group at Iowa State University to investigate the biosynthetic pathway of beta carotene in yellow maize in an attempt to determine what in the plant's tissues controlled or regulated the accumulation of beta carotene in the leaves and the endosperm. The researchers discovered that it was a certain gene known as *y1*. By 1990 they had isolated that gene and cloned a population of it into existence.

In their report on the project, the Iowa State scientists said, "An understanding of how the *y1* allele activates the carotenoid biosynthetic pathway in the endosperm of maize may suggest strategies for activating this pathway in the endosperm of other important grains such as rice or wheat."

Later, the lead author of that study, Brent Buckner, led a different group of researchers in a project to sequence the *y1* gene. They found that what it codes for is phytoene synthase, an enzyme involved in the biosynthesis of carotenoids, including beta carotene.

Separately, Gary Toenniessen also funded a team at the University of Liverpool to find out whether a precursor of beta carotene already exists in the rice endosperm. The team members, working under biochemist George Britton, discovered that there is indeed such a thing in the rice grain: a molecule with a longish chemical name abbreviated as GGPP.

This means that there exists in the rice endosperm a biochemical, GGPP, that is known to function as a raw material in the synthesis of beta carotene. The challenge was to get that synthesis going in the rice grain itself. One way of doing it, perhaps, is to put the maize *y1* gene into rice by means of genetic engineering. The idea is that the newly introduced *y1* gene would force the production of phytoene synthase in the rice endosperm. The phytoene synthase enzyme would in turn chemically alter or activate the GGPP that is already in the rice grain and start it on the pathway to expressing beta carotene. Ultimately, this would give you yellow rice.

And there, for the time being, is where that promising line of thought ended.

■ As it turned out, Gary Toenniessen and the research teams he had funded were not the only ones pursuing the idea of creating a rice whose grains bore beta carotene. In the late 1980s, at the Federal Institute of

Technology (ETH) in Zurich, a German scientist by the name of Ingo Potrykus was himself using molecular biotechnology techniques in an effort to improve global "food security," which meant providing people with a reliable access to sufficient quantities of food. This is a concept that Potrykus had ample familiarity with from personal experience. He was born in 1933, in Hirschberg, a mountain town in central Germany (the city is now Jelenia Góra, Poland), where his father was a physician in a local military hospital. In February 1945, when he was twelve, Potrykus and his family fled the town ahead of the advancing Russian army. His father died in the war's waning days.

"We had lost our home, our property, our father, and the financial basis of our life," Potrykus has said. "These were tough but valuable times; hunger was the predominant experience." Indeed, he and his two brothers often had to beg for—and sometimes even steal—food.

But he wound up going to college, and in 1968 got a PhD in plant genetics from the Max Planck Institute for Plant Breeding Research, in Cologne. There he was introduced to the concept of *totipotency* in plant cells: the ability of immature cells to give rise to other plant cells of any type. By the use of proper cell culture methods, it was even possible to generate an entire new plant from just a single totipotent cell. His idea was to alter the genetic makeup of single cells in such a way that would change the resulting plant for the better: to increase crop yields, for example, or make plants ready for harvesting earlier in their growth and maturation cycle.

By 1986, Potrykus had moved to ETH in Zurich, where he was directing a group of some 60 grad students, postdocs, and other colleagues working on the genetic engineering of crops. But in Switzerland, GMO research was very much a high-risk activity.

Traditional breeding techniques are simple to understand and execute. They are carried out on visible, tangible, and the most prosaic of macroscopic objects, plants. Even pollen grains are visible to the naked eye, and pollinating one plant by using pollen from another is so much like human sexual reproduction that botanists speak of male and female plants, as well as plant parts, such as ovaries. Everyone who knows about sex can understand conventional plant breeding perfectly well.

All of that changes when it comes to molecular methods that operate on a level where nothing is visible or tangible, and that deal with structures—genes and their components—with which few people are familiar, much less comfortable. For those reasons, many people view genetic engineering as a species of the black arts, an interference with the natural order of things, a blasphemy bordering on witchcraft, or worse.

That mind-set was and remains all too common in Europe, and it posed one of many obstacles that Potrykus and his colleague Peter Beyer had to overcome to create a provitamin A rice. For one thing, Potrykus was pursuing this research in one of the most violently anti-GMO countries on the Continent, Switzerland. Students there were so extremely hostile to biotechnology—at times shouting him down at public lectures—that he actually began to fear for his personal safety. In the dozens of popular articles that have been published about him over the years, one will occasionally read of Potrykus's "hand grenade–proof glasshouse" or "bomb-proof greenhouse" or something of the sort. Unfortunately, this is no exaggeration.

In 1987, when he was promoted to a full professorship at ETH Zurich, the university's administrators realized that special measures would have to be taken to ensure Potrykus's safety and that of his lab personnel, experimental plants, and the associated machinery and instrumentation necessary for transgenic plant research. The easiest solution would have been simply to build him a concrete bunker, but Potrykus was convinced that his plants required optimal growth conditions, including natural light, which meant that he needed a glass structure.

The greenhouse that resulted was essentially a *glass* bunker. It had 8-centimeter-thick glass walls arranged in five layers, the outermost of which was Panzerglas, a laminated armored glass that can withstand shell impact, bombardment, and blasting effects. Such glass is commonly used at bank counters, in jewelry shop windows, and in the windshields and doors of armored cars and trucks. Capable of resisting multiple bullet strikes from firearms, Panzerglas was in this case protecting humble rice plants. A second, inner layer of glass filtered out infrared rays to keep the interior from overheating.

All of that was to safeguard the lab workers and their experimental organisms. To protect the public at large from the possible escape of those same organisms—which many Swiss regarded as equivalent to radioactive materials, poison gas, or hazardous waste—the greenhouse had to be constructed to bacterial-tight biosafety standards. This meant airlocks, air filtration systems, biosafety cabinets and fume hoods, plus a disposal system in which every type of material leaving the greenhouse, including water, plants, and soil, had to be autoclaved to achieve a sterile level of decontamination and purity. The interior of the greenhouse was subdivided into 10 compartments, each of which could grow a different set of plant samples throughout the year, in an environment where the internal temperature was kept between 18 and 27 degrees Celsius (64 and 80 degrees Fahrenheit).

"Quite impressive and unique," Potrykus said of the completed structure.

Now, finally, he could turn his attention to rice. "Why rice? Because I wanted the greatest possible impact," Potrykus said. "And there was nothing more important than rice."

By 1990, Potrykus and a small band of his postdocs had already created the world's first transgenic indica rice. There are two principal types of rice consumed by humans: indica, grown in hot, wet environments common in lowland farms in Asia, and japonica, grown in more temperate, dry, upland areas. The trait that Potrykus and his group inserted into an indica strain has no food value but rather codes for resistance to an antibiotic, hygromycin, much as Cohen and Boyer had inserted a kanamycin resistance gene into *E. coli*. Such an antibiotic resistance gene when expressed in the plant constitutes a reporting system, as it tells you whether the transgene has been successfully inserted. It is therefore called a "marker gene." Potrykus's experiment showed that it was possible to introduce a marker gene into indica rice and to have that gene expressed not only in the primary transgenic plant but also in its progeny.

Having established that such a thing could be done, Potrykus's next goal was to work toward what he was really after: to make rice more nutritious by engineering the rice to accumulate beta carotene in the endosperm. This would require rewiring the rice genome . . . but just a little bit. The rice plant already expresses beta carotene everywhere else: in the leaves, stems, and even in the hull that encloses the seed, but the hull is removed during the milling process that produces white rice, taking all of its natural and native beta carotene with it. For some unknown reason, the expression of beta carotene was blocked or switched off in the endosperm. This meant that putting beta carotene into the grain was simply a matter of unblocking or switching on the biosynthetic pathway that already existed everywhere else in the plant, and letting the beta carotene be synthesized in the kernels as well.

This was another operation that is conceptually simple and easy to express in words. The challenge was implementing it in the lab, for Potrykus really didn't know how to do it or what the exact steps would be. And so, in the time-honored manner of senior professors at major educational institutions everywhere, he gave the problem to a grad student to work on as his PhD project.

The grad student was Peter Burkhardt, and he didn't know how to do it either. But he knew someone who probably did, and that was Peter Beyer, a cell biologist at the University of Freiburg, about 80 miles away from Zurich, in Germany. Beyer was a specialist in daffodils, a flower whose yellow color comes from beta carotene, making him an expert on the carotenoid biosynthetic pathway.

Daffodils, Beyer knew, contain a gene for phytoene synthase, the gene *psy*. It is a different gene from the *yl* gene that codes for the same protein in maize, but it performs the same function in daffodils: it codes for phytoene synthase, an enzyme that is active in beta carotene biosynthesis. It transforms the precursor molecule GGPP into a second, intermediate precursor, phytoene.

Beyer agreed to act as Peter Burkhardt's thesis advisor. To fund Burkhardt's project, which would require a massive amount of experimental lab work, Potrykus approached Nestlé, the world's biggest food company, but was turned down. At this juncture, Gary Toenniessen once again stepped into the biofortified rice picture.

In 1993, Toenniessen, who was by then director of the foundation's International Program on Rice Biotechnology, was sponsoring a workshop, "The Potential for Carotenoid Biosynthesis in Rice Endosperm," at the Rockefeller Foundation offices in New York. He invited to this event about 20 of the world's scientists who were knowledgeable about rice biochemistry, including Peter Burckhardt, Peter Beyer, and Ingo Potrykus. By this time, the trio had among themselves developed a scheme for putting beta carotene into rice.

In June 1993 the two Peters and Potrykus traveled to New York aboard the same plane.

The Rockefeller workshop took place over two days, and by the end of it the scientists had agreed there was a possible strategy for establishing a beta carotene biosynthetic pathway in the rice endosperm. The scheme, as advanced by Beyer and Potrykus, was to introduce all of the enzymes needed to activate the precursors (or their products) to beta carotene that already existed (or could be produced) in the rice kernel. This would require four separate enzymes, the product of four separate genes, and they all would have to function together serially, one after another. Potrykus's idea was to introduce the four genes into rice by inserting one gene into each of four rice plants independently, and then to bring them all together by crossing. This was perhaps a slightly Rube Goldberg, roundabout way of getting four separate genes into one and the same plant, but it was at least theoretically possible. Maybe.

"Many of the participants thought that such a project did not have much chance of success," Ingo Potrykus said afterward. "There were hundreds of scientific reasons why the introduction and coordinated function of these enzymes could not be expected to work."

"I think all of us recognized that this was a high-risk project," Gary Toenniessen said. "Nobody had previously introduced four genes into a plant and gotten them to function in a sequential way that allowed a whole biosynthetic pathway to be introduced, so it was cutting-edge research."

Peter Beyer was himself so skeptical that he privately thought the plan would probably fail.

■ But the Rockefeller Foundation was gung-ho for cutting-edge research, no matter how chancy it was, so long as there could be a really big payoff at the end, as there would be if the Potrykus-Beyer scheme actually worked. So the foundation was willing to fund their project and ended up supporting both Beyer and Potrykus, and the work of their respective teams, which included Peter Burkhardt.

With two fully equipped and complementary labs at their disposal—one in Zurich and the other in Freiburg—Potrykus, Beyer, and Burkhardt now confronted the biggest obstacle of all: actually rewiring the genes of a rice plant. This involved identifying, isolating, and then inserting into the plant's genome a series of genetic sequences that would create a new biochemical pathway, a series of metabolic reactions, that would in the end cause the rice to produce beta carotene in its grains.

Ambitious as it was, accomplishing that goal would not require anything like a total, ground-up rewrite of the rice plant's genetics. As we have seen, rice already expresses beta carotene, it is practically bursting with the stuff in its leaves, stems, and all of its green tissues. Carotenoids, of which beta carotene is one type among more than 600, occur in all photosynthetic plants (and even in some bacteria), and are essential factors in photosynthesis. In *Oryza sativa*, this means that all of the metabolic machinery and raw materials for producing beta carotene are already inside the plant to begin with, where they are fully active and functional. The challenge was to get them active and functional in the endosperm as well.

The primary raw material and starting point of the full biosynthetic reaction that leads to beta carotene is the molecule GGPP, whose full chemical name is geranylgeranyl diphosphate. This is a 20-carbon-long molecule, symbolized as C_{20}.

Now, just as beta carotene is a vitamin A precursor, GGPP is itself a precursor of beta carotene. To get from it to beta carotene requires four enzymatic steps, the first of which is to combine two GGPP molecules into one large 40-carbon-long molecule, phytoene (C_{40}), which is a colorless carotene. That reaction is under the control of the enzyme phytoene synthase (PSY), which condenses the two shorter molecules into the longer one, phytoene.

Schematically:

GGPP (C_{20}) + phytoene synthase (PSY) → phytoene (C_{40})

From his work on daffodils, Peter Beyer knew that a specific type of daffodil (*Narcissus pseudonarcissus*) has a gene that codes for the needed enzyme, phytoene synthase. This is the *psy* gene,[*] and its structure was precisely known. It is exactly 1,548 base-pairs long in messenger RNA (mRNA). This information, and the base-pair sequence itself, were on file under a unique identifier code (accession number X78814), at several genome databases, including the European Molecular Biology Laboratory, and at GenBank in the United States. In fact, it was Beyer himself, along with other members of his lab, who put it in those data banks in 1995.

The first thing that Potrykus, Beyer, and Burkhardt had to do, then, was to incorporate the *psy* gene into the rice genome. In molecular biology, genes are not introduced bare, by themselves, but must be flanked by a promoter sequence, which acts to initiate the reaction, and some other biochemical helper molecules. The entire construct is then ready to be inserted.

The next step was to insert it. For all the sophistication of their work, and the extreme delicacy of the molecular materials they were working with, what they actually used to place the *psy* gene construct into the rice genome inside the tiny cells was something quite primitive. Amusing, in fact; almost comical.

They used a gun.

■ Molecular biologists knew that it was difficult to insert exogenous DNA sequences into plant cells. Plant cells are protected by relatively strong cell walls, with a dense meshwork of cellulose fibers that prevent foreign DNA from entering the cells. One way around this barrier is to strip off the cell wall, leaving a "naked" cell called a protoplast, into which genes can

---

[*] It is a scientific convention to abbreviate gene names in lower case italics (*psy*), and the corresponding protein that the gene coded for in upper case roman type (PSY).

be introduced fairly easily. Ingo Potrykus had tried that technique himself, and indeed his first transgenic, antibiotic rice was produced in this way. But this is a messy, labor-intensive process, and it turned out that in actual practice simply "shooting" DNA-coated microparticles into rice embryos was just as effective and required less work.

The mechanism that did this was called a "gene gun," which literally shoots DNA into cells. You simply load the gun with particles coated with genes, point the gun at a leaf or other plant tissue containing the cells you want to transform, and pull the trigger, just like that. One might think that such treatment would kill the leaf, or the cells, but it doesn't. Instead, it gives the cells inside it a new kind of life—at least those that survive.

Also called a *bioballistic* or *biolistic* particle delivery system, the device was invented by two biologists (John Sanford and Ted Klein) and two electrical engineers (Edward Wolf and Nelson Allen), all of Cornell University. During the winter break of 1983, Sanford and Wolf purchased a Crosman air rifle, loaded it with tiny, micron-sized tungsten particles coated with a DNA marker gene, and fired the particles into onion cells, which were large and presented easy targets. The surprise was that this bullet-hole process actually worked, as was proven when the transformed onion cells grew out and duly expressed the inserted marker gene. Amazing as it was, you could actually do genetic engineering by means of a BB gun.

The inventors later produced a variety of better-looking and more sophisticated devices that also used high-velocity microparticles to deliver nucleic acid sequences into cells. They filed for a patent on their device and in 1987 published an account of how it worked in the journal *Particulate Science and Technology*.

"The concept of particle bombardment has been put forward as a universal mechanism for transporting substances into any living cell," the inventors wrote. "An acceleration device has been designed and constructed which can accelerate small tungsten particles (1 to 4 μm in diameter) to velocities of about 1,000 to 2,000 ft/sec. We have found that these particles can penetrate cell walls and membranes and enter cells in a nonlethal manner. Thousands of cells can be penetrated simultaneously, in situ, as they occur in tissues. Particle bombardment has been shown to be effective in delivering foreign substances into a variety of plant species, including onion, tobacco, corn, and rice."

And so when Potrykus, Beyer, and Burckhardt first tried to put the *psy* gene into rice, they used the gene gun simply because it was known to work.

The whole point of injecting the *psy* (phytoene synthase) gene into the rice genome was to see whether it would activate the C_{20} GGPP molecules and get them to condense into the larger C_{40} molecule, phytoene. This was an important step, because phytoene is itself a carotenoid, and to have it expressed in the rice endosperm would mean that carotenoid synthesis was possible there. Conversion of GGPP to phytoene was known as the first committed step in the synthesis of carotenoids in plants.

In the spring of 1993, the team members used a particle inflow gun to bombard a total of 590 specially prepared Taipei 309 rice embryos with a blast of 1- to 3-micron gold particles coated with their *psy*-gene and promoter constructs. It took a while to get results, and it wasn't until the end of 1994 that they had their first successful transformations. From the original batch of 590 embryos, the team members regenerated 87 plant lines, 47 of which were fertile, and 27 of which contained an intact *psy* gene. The artificially introduced *psy* gene had been successfully put into the rice genome and had been properly expressed in rice plants.

The scientists analyzed the endosperms of those plants by means of high-performance liquid chromatography, which was a technique used in analytic chemistry to separate, identify, and quantify each component in a complex mixture of chemicals. And when they did they found that phytoene, a carotenoid, was indeed present.

The team members, principally Peter Burkhardt, Peter Beyer, and Ingo Potrykus, reported their results in a 1997 *Plant Journal* paper that became one of the foundational documents in the history of Golden Rice.

"Phytoene was unequivocally identified in several transgenic rice endosperms," the authors wrote. "Thus, it is demonstrated for the first time, that it is in principle possible to engineer a critical step in provitamin A biosynthesis in a non-photosynthetic, carotenoid-lacking plant tissue. These results have important implications for long-term prospects of overcoming worldwide vitamin A deficiency."

But, as significant as it was, that critical first step alone was still a long way from the Golden Rice prototype.

■ In fact, it would take the researchers almost five more years to produce rice grains that were visibly yellow in hue. To get to that point, the scientists needed to add three additional enzymes to the rice cell. Two enzymes would be required to produce a second intermediate substance, lycopene, while a third would be necessary for synthesizing the final end product, beta carotene, from lycopene.

The scientists encountered several problems in their attempts to engineer the remaining steps. For more than a year, Peter Burckhardt worked on putting the next gene in sequence (phytoene desaturase) into the rice genome, which he did, but he was able to get only heavily distorted transgenic plants out of the process. Ultimately, the team did manage to get all four separate genes into four separate plant lines, but once again the desired result eluded them.

"When we finally had transgenic plants for all genes separately, we were able to combine genes pairwise, but all this did not look too promising," Ingo Potrykus said. "We had used biolistic transformation of embryogenic suspensions and precultured immature embryos but had the typical complex integration patterns and unfit plants, and a hybridized combination caused problems with gene stability and plant fertility."

It was now 1998, six years into the project, and the group had essentially reached a dead end. Plainly, a substantial midcourse correction was called for, and so the scientists ended up making three big changes. First, they added two new members to the research team. One was Xudong Ye, who was from China and was just then finishing up his PhD work on transgenic ryegrass at ETH. The other was Salim Al-Babili, who was from Syria and had completed his doctoral dissertation under Beyer's direction in Freiburg. After considerable brainstorming, the new group opted to make the second big change, which was to abandon the gene gun approach in favor of another gene transfer technique that made use of a bacterium, *Agrobacterium tumefaciens,* as the gene vector. Third, instead of putting the genes into four separate plant lines and then combining them into one by crossing, they decided to insert all the necessary genes together, in one fell swoop.

The virtue of the *Agrobacterium tumefaciens* bacterium is that it mimics the natural process of bacterial conjugation: the bacterium transfers into the target cell a piece of DNA that achieves a certain result. In this case, that stretch of DNA would contain all the genetic sequences necessary to go all the way from GGPP to beta carotene. These included the daffodil *psy* (phytoene synthase) gene, a second gene, *crtI* (phytoene desaturase), from the bacterium *Erwinia euredovora,* and two more that produced the final needed enzymes: ζ-carotene desaturase and lycopene ß-cyclase.* If everything went according to plan, the result would be the miracle molecule and grand prize, beta carotene.

When the new gene constructs had been prepared, the scientists put them into 500 immature rice embryos, from which they got a number of transgenic plants that were of a normal phenotype and yielded fertile offspring. They then transferred 10 of these plants into the greenhouse for setting seeds.

About four months passed before the plants had grown to maturity and produced actual rice grains. Peter Beyer brought them to his Freiburg lab for polishing, after which he would know whether the endosperm inside the hulls were white or yellow. Beyer polished the seeds by machine, a process that took about six hours.

Early one evening in February 1999 he was finished, and he placed a phone call to Potrykus. "Ingo, open your computer," he said. "I am sending you a picture you will enjoy."

There on the screen in front of him were about 100 grains of rice. And they were yellow.

---

* Other elements in the final genetic constructs included a promoter sequence, an antibiotic-resistance marker gene, and miscellaneous other molecular auxiliary structures.

"I do not recall how many polished endosperms we had looked at over the years," Potrykus said later. "And now there were some golden ones . . . the first golden endosperms."

They looked like gems, he thought. But to him those yellow grains of rice were much more precious than jewels.

3

GR 0.5 AND BEYOND

I t had taken Peter Beyer and Ingo Potrykus about six years to invent the rice that Peter Jennings, back at the International Rice Research Institute in April 1984, had spoken of as "a rice with yellow endosperm." What the inventors now had in front of them was exactly that, and the yellow was not mere food coloring, it was demonstrably beta carotene, as shown by high-performance liquid chromatography tests. The inventors had also determined how much total carotenoid was contained in the grains: about 1.6 micrograms per gram of endosperm (1.6 μg/g).

It was a relatively small amount, but this was only a prototype, a barebones, skeletal laboratory specimen—a proof of concept, not a mass-production version ready for release. To make the food into a useful tool in combating vitamin A deficiency, the endosperm would have to contain more carotenoid, because beta carotene constituted only a part of the total carotenoid in the grains. In addition, some of that amount would be lost while the rice was in storage, and even more during the process of cooking. To be really useful, the rice would have to contain at least 2 micrograms of beta carotene per gram of grain (2 μg/g), which would correspond to 100 micrograms of retinol if a daily 300-gram (10-ounce) dry-weight portion of the rice were consumed. The experimenters thought, however, that this was a realistic goal; it would just require some further development.

In the spring of 1999, the scientific team wrote up its results and submitted the paper to the respected British science journal *Nature*, with a cover letter explaining that what the researchers had accomplished was not only an important advance in itself but was also a contribution to the wider public debate about GMOs.

"The *Nature* editor did not even consider it worth showing the man-uscript to a referee, and sent it back immediately," Potrykus said. "Even supportive letters from famous European scientists did not help. From other publications in *Nature* at that time we got the impression that *Nature* was more interested in cases which would rather question instead of support the value of genetic engineering technology."

After that surprising rebuff, the authors then sent the paper to the American journal *Science*, which duly published it in January 2000. The magazine had initially announced the rice earlier, in a brief news piece ("New Genes Boost Rice Nutrients") in August 1999, adding an editorial comment, "This achievement is a Herculean feat of gene engineering."

On March 3, 2000, Beyer and Potrykus filed for a patent not on the rice per se but on a "method for improving the agronomic and nutri-tional value of plants." But it was in July 2000 that Ingo Potrykus became world famous. That was when the American edition of *Time* magazine ran its cover story on Golden Rice, with Potrykus front and center, sur-rounded by his green, leafy rice plants, with the caption "This rice could save a million kids a year." His mass media debut was followed by satu-ration coverage on television, radio, and in the international press.

The product by now had officially been dubbed "Golden Rice." The name had not originated with Ingo Potrykus or Peter Beyer but had come from a somewhat unlikely source. In September 1999, the Rocke-feller Foundation's Gary Toenniessen and some of his colleagues were having dinner in Bangkok with Mechai Viravaidya, the head of an NGO in Thailand. Viravaidya was very savvy about advertising, marketing, and promotion, and when he heard Toenniessen and the others talking about "yellow endosperm rice," he was aghast. "Don't call it 'yellow en-dosperm rice,'" he blurted out. "Call it *Goooolden* Rice. You've got to have a marketing campaign behind this. You got to make it a *treat* to eat Golden Rice. It's got to be better than *white* rice."

"I remember going back and telling Ingo that we had to start calling his invention 'Golden Rice,'" Toenniessen said later. "Ingo caught on, and the rest is history."

At this point, then, with a valuable product in hand and even an at
tractive name for it, Beyer and Potrykus could be forgiven for imagining
that the hard part of their project—which they had hitherto thought was
the scientific challenge of getting provitamin A into the rice endosperm—
was behind them. It should be a relatively easy matter, they thought, to
boost beta carotene levels in the endosperm and then to get an improved
version of the rice (GR 1) out of the lab and into the hands of the small-
scale farmers for whom it had always been intended. But in this they
couldn't have been more wrong.

■ First there was the matter of intellectual and technical (or "tangible")
property rights: IP and TP, so called. It is one thing to utilize patented
technologies, techniques, and processes with abandon while in the
course of active research. That is standard practice. Experimenters often freely share their own private knowledge, procedures, materials,
and tricks with other researchers. They might send samples of reagents
they had developed, enzymes, even organisms, sometimes placing certain use restrictions on them, other times not. This is a matter of professional courtesy among those working within the same area of applied
science.

But it is another thing entirely when the results of that research become, or threaten to become, an actual, saleable product. Even if it were
to be given away for free, the rights to any IP or TP that might be attached to or embodied in the product would have to be respected.
Knowing this, and in the expectation that Golden Rice would soon be
released for use by farmers, the International Rice Research Institute in
the Philippines asked the Rockefeller Foundation to undertake an IP
and TP audit of the technologies and procedures that had been used in
the rice's development. The foundation agreed but realized that, since
this would be a monumental effort, it would be most efficient to contract
out the task, which was performed by a nonprofit biotechnology clear-
inghouse at Cornell University, the International Service for the Acquisition of Agri-biotech Applications (ISAAA).

A group of ISAAA investigators reviewed the numerous technologies, processes, substances, and organisms that were relevant to the development of Golden Rice. These included plant and seed sources, genes, gene constructs, cloning vectors, enzymes, germplasms, plasmids, plant regeneration techniques, and so on. They discovered patent holders as diverse as DuPont, Eli Lilly, and Hoffman–La Roche; Rutgers, Stanford, and Columbia Universities; individual inventors; and Kirin Brewery Co. Ltd., of Japan (the company made beer from rice). There were licenses, patents, and other encumbrances attached to Bio-Rad's microprojectile bombardment apparatus, Monsanto's promoters and terminators, and even the arcane "small subunit transit peptide," which was part of the construct.

"The deconstruction of GoldenRice™ was complex," the investigators wrote in their 56-page report on the audit. "It yielded over fifteen tangible property components and approximately seventy patents and related IP that seem to have been integral to the product's development."

Product clearance—the process of acquiring rights or licenses to all the IP and TP used in the invention—was stacking up to be a major headache, especially in view of the fact that the rice was intended to be distributed in many countries whose patent laws were different from those in the United States. Plus there were the further facts that new patents are always being filed, old ones expire, companies merge, split up, get bought and sold, go bankrupt, etcetera. To complicate matters even further, there was just at that time a four-way patent litigation in progress involving the *Agrobacterium* transformation system that they had used to make the product, "and precise ownership cannot be established at this stage."

In short, what a mess. Virtually every step in the creation of Golden Rice suddenly loomed as a nightmare of legal intractability.

After reading the audit report, Ingo Potrykus emailed the lead author, "Your analysis has led to a frightening picture of the future: how should one be able to achieve freedom to operate for the golden rice if 32 patent holders have to be asked to release their patent rights for the humanitarian project for 78 rice-growing countries."

With visions of his precious Golden Rice evaporating into little more than a genetic engineering curiosity, Potrykus's first instinct was to join the crusade against patenting. He realized soon enough, however, that all these patents had actually *facilitated* the development of Golden Rice. After all, the technologies that he and Beyer had used had been publicly available only because the inventors of the technologies knew that their rights were protected and secure. If that were not the case, then many of the techniques and processes that he and Beyer had used would most probably have remained proprietary: corporate or personal secrets.

There was a way out of the IP/TP morass, but wending a clear pathway through the maze was going to be a tricky business.

But before that process could even begin, there was an important prior matter to be dealt with. It so happened that a portion of the funding for the Golden Rice research had come from the European Commission, which had supported Peter Beyer's work as part of the European Community Biotech Program and its Carotene Plus project. Participation in that program was conditional on making a European commercial firm a partner to the research, which would have rights to any product resulting from it. To satisfy this requirement, Beyer and Potrykus now made a number of hocus-pocus legal moves in quick succession.

First, on February 20, 2000, they assigned all rights to their technology to Greenovation, a small licensing company affiliated with the University of Freiburg. Then, on April 14, 2000, Greenovation transferred rights to the same technology to Zeneca Group PLC, a British pharmaceutical company with headquarters in London. Under the terms of the rights transfer, Zeneca would acquire commercial rights to the product while the inventors would retain rights to the humanitarian use of the rice. And, on that same day, April 14, Zeneca also granted licenses back to the inventors so that they could make their technology available, free of charge, to resource-poor farmers in developing countries.

By this series of maneuvers, Beyer and Potrykus satisfied the European Commission's requirement that a European commercial firm (Zeneca) be a partner to the research while it simultaneously left intact Beyer and Potrykus's commitment to distribute their technology to small-scale farmers in developing countries at no cost. For purposes of the agreement with Zeneca, a small-scale farmer was defined as one whose annual sales did not exceed US$10,000.

Still, none of those actions addressed the broader IP/TP problem. It was finally solved by an individual who can arguably be regarded as the savior, rescuer, or deliverer of Golden Rice from the intellectual and technical property rights quagmire, Adrian Dubock.

Dubock, a direct and assertive man of seemingly inexhaustible energy, drive, and capacity for hard work, was born in the United Kingdom. In 1976 he acquired a PhD from Reading University with a thesis on the reproductive physiology and ecology of the grey squirrel, after which he took a job with the Ministry of Agriculture, Fisheries and Food.

By 2001 Dubock was global head of mergers and acquisitions, ventures, and IP licensing at a new company called Syngenta. Syngenta was the corporate entity that Zeneca had become after it merged with Novartis, a Swiss drug firm. Dubock was an enthusiastic supporter of Golden Rice, practically a messiah, and he was living in Switzerland, within striking distance of both Beyer and Potrykus. When he undertook an IP/TP survey of his own, Dubock discovered that matters were actually not quite as hopeless as they had appeared to the ISAAA patent investigators at Cornell. For one thing, there was a difference between the technologies and processes that the experimenters had used during their six-year-long course of research and those that the final product actually embodied, which amounted to just a few. For another, Monsanto, without even having been asked, had already donated its IP rights in the product to Beyer and Potrykus. ("A really amazing quick reaction of the PR department," Potrykus thought.)

Within the space of six months, Dubock negotiated deals with the remaining patent holders such that Beyer and Potrykus could make use of all the protected technologies contained in their genetically engineered rice plant.

And so at this point, having now acquired complete legal "freedom to operate," the inventors once again imagined that they were over their last big hurdle to the distribution and use of Golden Rice. Far from it!

■ In January 2001, Ingo Potrykus, Peter Beyer, and Adrian Dubock, the trio who would oversee Golden Rice development from then on, hand-delivered the first shipment of their yellow endosperm rice seeds to IRRI in the Philippines. The kernels had been grown in the ETH bomb-proof greenhouse, and in all they amounted to only about three ounces of rice—some 600 seeds—enough to fill the double-walled plastic bag that Potrykus had carried in his pants pocket on the flight to Manila. In addition, they carried with them six 2.5-centimeter tubes of the genes that would turn any type of white rice into a Golden Rice factory. The three men were met at the airport by an IRRI representative, and Potrykus handed over the seeds and the genes at the customs station.

It was a rather commonplace method of delivery of what was potentially a historic agricultural entity. Three ounces of seed might not sound like a lot of rice, but in fact only a single one was enough to start a food revolution.

As Ingo Potrykus liked to say, and indeed as bears repeating, "The entire technology is embedded in the seed. All a farmer needs to benefit from this technology is one seed. Each seed can produce 20,000 metric tons of Golden Rice within two years."

The plan was that the IRRI rice breeders would plant these new seeds and then *introgress* (backcross) the germinating Golden Rice plants with local rice varieties to produce types that would grow well in the tropics. Based on long experience, Potrykus estimated that fewer than eight generations of backcrossing would be needed to introduce the beta carotene trait into any given local rice type. Everyone was optimistic, and IRRI rice scientist Gurdev Khush said that he expected Golden Rice to be in farmers' fields in Asia by 2002. This was the basis for the oft-repeated claim that Golden Rice was "ready" in 2002. (It wasn't.)

Trouble began practically the moment that Golden Rice crossed the border into the Philippines, for this was a development that set off alarm bells among the ranks of anti-GMO activists. On February 9, 2001, barely a month after those three ounces of Golden Rice arrived at IRRI, Greenpeace, which had an office in Manila, issued a press release dismissing Golden Rice as "fool's gold," called it "a dangerous technology" (while offering no evidence that it was even remotely risky), and claimed that it "will not solve the problem of malnutrition in developing countries" (which is never what it was intended to do in the first place). Syngenta, for its part, came across as a greedy and deceitful corporate monster.

"Syngenta, one of the world's leading genetic engineering companies and pesticide producers, which owns many patents on the 'Golden Rice,' blatantly boasts that a single month of marketing delay of 'Golden Rice' would cause 50,000 children to go blind," the press release said. "Greenpeace calculations show however, that an adult would have to eat at least 3.7 kilos [8 pounds] of dry weight rice (i.e. around 9 kilos [20 pounds] of cooked rice), to satisfy his/her daily need of vitamin A from 'Golden Rice.' . . . A breast-feeding woman would have to eat at least 6.3 kilos [14 pounds] in dry weight, which converts to nearly 18 kilos [40 pounds] of cooked rice per day."

As we have seen, the bioavailability of beta carotene in Golden Rice had not yet been established, so there was no scientific basis for such claims. Nevertheless, in a March 4, 2001, opinion piece ("The Great Yellow Hype") in the *New York Times*, food writer Michael Pollan echoed Greenpeace's astonishing figures, saying that "an 11-year-old would have to eat 15 pounds of cooked golden rice a day—quite a bowlful—to satisfy his minimum daily requirement of vitamin A."

Given this ridicule and repudiation of Golden Rice, it was incongruous that on the same day of the press release's issuance, February 9, 2001, Benedikt Haerlin, Greenpeace International's "genetic engineering campaign coordinator," stated at a press conference in Lyon, France, that, because of claims that it could save lives, Golden Rice posed a moral dilemma for Greenpeace. He said further that the organization would not attack any GMO rice field trials in the Philippines. This ran against the official policy of Greenpeace, which was intransigently and unalterably opposed to genetically modified organisms and had a proven track record of destroying fields of such crops—and of getting away with it. Indeed, in September 2000, 28 Greenpeace anti-GMO campaigners were acquitted of destroying six acres of GM maize in Norfolk, England, despite the fact that they openly admitted in court to having done so.

But a week after Haerlin's press conference, he wrote a letter to the editor of the London *Independent* in which he clarified his position in a way that made it more consistent with Greenpeace doctrine, history, and current (and future) practice: "The acknowledgement that we have no plans to disrupt field trials of golden rice is in no way a U-turn in our policy, but simply practical recognition of the resources we have available for our work in the Philippines. Golden rice has not been ruled out as a target for direct action in the future."

This was the beginning of an image problem that would beset Golden Rice and its champions henceforth: a sweeping Greenpeace denunciation and public relations campaign against it, coupled with threats of "direct action." Such threats would haunt Ingo Potrykus for years into the future.

■ Although by this point public sentiment in Europe was generally hostile to GMOs, there had been a brief period when genetically modified foods were regarded as a nonissue, or even as acceptable. And, in the United States, at least one GMO food was for a time wildly popular.

The world's first commercially marketed, genetically modified food was an "improved" version of the common tomato. The product was the Flavr Savr tomato, developed in 1988 by Calgene, a small company in Davis, California, that at one stage during the prototype's development had just three employees.

The new tomato's raison d'être was better flavor coupled with longer shelf life. Normally, tomatoes are picked, boxed, and shipped while still green, because naturally ripened tomatoes tend to soften and spoil during transport. The green tomatoes are artificially ripened in warehouses with ethylene gas, an agent that is naturally given off by ripening fruit. But this gassing process comes with a price: although the resulting tomatoes look red and ripe, they nevertheless still taste green ("like cardboard"), because they lack the additional flavors produced by vine ripening. The Flavr Savr solution was to let the tomatoes ripen on the vine but to prevent them from softening by means of genetic engineering.

Compared to what was necessary to invent Golden Rice, the amount and nature of the genetic engineering required to create the Flavr Savr tomato was rather modest. Ordinary tomatoes are subject to softening when ripe because an enzyme called polygalacturonase (PG) degrades a stiffening agent (pectin) in the cell walls. Calgene's idea was that if you prevented the tomato's expression of the PG enzyme, then this natural softening process would not occur. The modified tomatoes could therefore be left to ripen on the vine, developing their full flavor, and then be shipped to consumers while red, ripe, and firm. With this "new and improved" miracle tomato, supposedly, you could make a fortune.

It was easy enough to disable the gene that coded for the PG enzyme. You clone the gene, manipulate it so that it reads backward (in reverse orientation), turning it into an *antisense* gene, and then you insert the modified gene back into the tomato genome by means of the dependable *Agrobacterium* vector. This would be a tiny, minimalist alteration to the tomato genome. But, because the PG gene's nucleotides are now strung across the genome in reverse sequence, the antisense gene is unreadable to the ribosomes (organelles that translate RNA into peptide sequences and proteins) when they try to synthesize what it coded for, which after the insertion of the antisense gene is nothing at all. The result is that the tomato fails to express the PG enzyme (or expresses less of it).

The plan actually worked when implemented, and the Calgene team filed for patents on its product and then published an account of its work in *Proceedings of the National Academy of Sciences*, in December 1988, under the title, "Reduction of Polygalacturonase Activity in Tomato Fruit by Antisense RNA." Although PG enzyme production was substantially reduced, lycopene, the carotenoid that gives the red color to tomatoes, continued to be produced normally.

Initial rounds of testing showed that the "improved" tomato was neither firmer nor better tasting than unimproved tomatoes. Crossing the original plant with different varieties, however, achieved better results. The first toxicity tests, by contrast, were somewhat alarming, because some of the rats that had been fed Flavr Savr tomatoes had developed stomach lesions. However, further rounds of testing disclosed that the same lesions were found in rats fed ordinary tomatoes, and even in rats not fed any tomatoes at all.

The FDA formally approved the product for sale on May 18, 1994, and it was on supermarket shelves three days later. "They sold like hotcakes," said Belinda Martineau, who was involved in the research, development, and testing, and wrote a first-person account of the project, *First Fruit* (2001). Although there had been some protests and demonstrations (and a false but persistent claim that the tomato contained a flounder gene), Flavr Savr tomatoes were for a time in such great demand that some stores limited their customers to two of the GMO tomatoes a day. There was even a waiting list of stores that wanted to sell them.

But, as it turned out, the costs of delivering the tomatoes to market were consistently higher than the selling price, and the company stopped selling the product in 1997. Thus, the first GMO food ever to be marketed was a technological and initial sales success but in the end a commercial flop.

All of this happened in the space of nine years—everything from the development of the prototype in 1988 and its withdrawal from the market in 1997. In other words, the product had been developed, tested, and had come and gone before Ingo Potrykus and company had their proof-of-concept Golden Rice in hand. And, even when they did, it would take many more years, and several more iterations of the events that had produced it, before a version was ready for real-world use.

■ At IRRI in the Philippines, the head breeder, Gurdev Khush, received the rice that Potrykus had brought with him from the ETH greenhouse in Zurich. Khush was a legendary figure in the rice research community, and during his time at IRRI, which spanned 30 years, he had developed more than 300 new varieties including IR36, which became one of the world's most widely planted rice types.

The Golden Rice that Potrykus and Beyer had invented was a japonica cultivar, Taipei 309. The Taipei 309 line was easy to work with in the lab and grew well enough in greenhouses, but it was not the most suitable type for growing in real-world conditions, which is to say, in the open fields of the tropics. Rather, it was indica rice varieties that were most widely grown and consumed in countries where vitamin A deficiency was prevalent. And so a team of IRRI rice breeders now set about the task of backcrossing the prototype Taipei 309 Golden Rice with indica types. The specific strain they chose to work with was IR64, which had been created in the 1960s by IRRI breeders and was now a locally important rice widely grown in Southeast Asia. This was but the first stage in a more extensive program of diversified variety development.

Growing those varieties, however, did not mean that the scientists were getting close to putting Golden Rice into the hands of farmers. For one thing, Golden Rice was still only a prototype, essentially a creature of the laboratory, an experimental plant. As it turned out, the prototype was merely the first of four Golden Rices that two separate groups of scientists would ultimately bring into existence: the public-sector group (Potrykus, Beyer, and their colleagues at Zurich, Freiburg, and elsewhere), and the private-sector group (Syngenta scientists). The follow-on versions, to be developed by these two groups of scientists in the years to come, would differ from the prototype in two major ways: they incorporated transgenes that would be more acceptable to regulators, and, with one exception, they produced higher levels of beta carotene in the kernels.

The nomenclature used in the growing literature about Golden Rice, and specifically the terms used to describe the different versions of it, has led to considerable confusion over the years as to which rice is which. The inventors used the terms "prototype" or "proof of concept" to designate the original Golden Rice that had been created in 1999 and was announced in the pages of *Science* in January 2000. They also used the term Golden Rice 1, or SGR1 (Syngenta Golden Rice 1), to refer to a follow-on, nutritionally improved rice developed by Syngenta scientists in 2003. That rice contained higher levels of total carotenoids in its kernels, as much as 6 micrograms per gram. However, other published accounts have erroneously referred to the *prototype* version as GR 1.

For purposes of clarity in keeping these different versions of Golden Rice straight, I adopt the common convention used in the computer software industry to distinguish among these four rices by numerical version identifiers. Thus, the original prototype, proof-of-concept Golden Rice is Golden Rice 0.0, or GR 0.0. A second version was also created by the public sector scientists. This version was not designed to improve beta carotene levels. Instead, it was designed specifically to make Golden Rice "amenable to deregulation." Deregulation is the process that leads to the registration of the product in a given country and the granting of permission for commercial or noncommercial use. To make the rice amenable to deregulation, the public-sector scientists made certain technical changes to the inserted gene construct. I refer to this second version, produced in the two-year period after the debut of the prototype, as GR 0.5. (Despite the higher version number, GR 0.5 rice types actually produced *less* total carotenoid content than the prototype.)

Two additional versions, created by Syngenta scientists and using different transgene constructs, came afterward. One of them did in fact have improved carotenoid expression levels in the endosperm: as much as 6 micrograms per gram, compared to the prototype's 1.6. This is Golden Rice 1, or GR 1. In 2004, finally, Syngenta scientists developed a new and even better version of the rice, one that had as much as 37 micrograms per gram average total carotenoids in the kernels, fully 23 times more than the prototype rice had contained. This was Golden Rice 2, or GR 2. But that would come in the future, fully five years after the appearance of GR 0.0.

■ Golden Rice 0.5 was created by Potrykus, Beyer, and other members of their respective teams, including a Vietnamese scientist named Tran Thi Cuc Hoa. Her name has come to be associated with this version of the rice because of her prominent role in its development.

The prototype Golden Rice had included a selectable marker gene that made the resulting plants resistant to the antibiotic hygromycin. However, some GMO critics had suggested that eating a plant food containing antibiotic resistance genes might lead to the unintended consequence of creating antibiotic resistance elsewhere. The idea was not that such foods would make *people* resistant to the antibiotic in question but rather that the gene might lead to resistance among the bacteria in the consumer's intestinal tract. Such an artificially induced resistance, these critics said, could then spread to the disease-causing bacteria themselves, by a process known as horizontal gene transfer, making those harmful bacteria immune to the antibiotics that doctors depended upon to cure diseases.

Such a scenario is highly implausible for several reasons. One is that for all of this to happen, the antibiotic resistance gene would have to escape the plant cell and then manage to survive the gene-breakdown substances normally present in saliva and stomach acids, which is unlikely. In addition, the gene would then have to establish itself in the right place in the genome of the disease-causing bacterium, which is also improbable. But in the special, closed world of GMO regulatory systems, any possible unintended negative outcome, no matter how remote or unrealistic, would have to be guarded against in the name of caution (or overcaution). Warranted or not, antibiotic resistance marker systems amount to red flags in the eyes of regulators, making it prudent to remove them from the picture entirely and replace them with something else.

And so in their attempt to create a Golden Rice line that would be more acceptable to regulators, Hoa and her team used another marker system, one that did not make use of an antibiotic resistance gene. Instead, they used one based on mannose, a kind of sugar related to glucose. Ordinary, untransformed cells can use ordinary sugars—sucrose or glucose—as energy sources. Genetically modified cells, however, can be programmed to utilize a different sugar, mannose. Such cells would survive in a culture medium containing only mannose, while untransformed, ordinary cells would die off. Thus, the Golden Rice scientists inserted the gene *pmi*, which codes for phosphomannose-isomerase, a form of the new sugar, yielding an antibiotic resistance–free marker system.

With that change, plus several other improvements to the inserted gene construct, but retaining the same daffodil *psy* gene that had been the basis for GR 0.0, the researchers attempted to genetically engineer the Golden Rice pathway into two indica cultivars that were popular in Southeast Asia: IR64 rice from the Philippines, and MTL250 rice, which was popular in Vietnam. They also introduced the new transgene into the same japonica variety, Taipei 309, that Potrykus and Beyer had used in the prototype.

The researchers were successful in replacing the antibiotic selectable marker with the mannose-based version and produced in the second generation seeds that were of a definite yellow-golden hue. The group members published their results in 2003, under the title "Golden Indica and Japonica Rice Lines Amenable to Deregulation." But although the new GR 0.5 varieties might now be more acceptable to regulators, these same varieties were less successful nutritionally, containing lower levels of beta carotene in the endosperm than the original, proof-of-concept Golden Rice.

Still, there was a silver lining to Golden Rice 0.5 in the form of a novel biochemical that the scientists observed in the endosperm. It was "a healthy antioxidant compound, claimed to be more efficient than vitamin E." This, the authors said, was an "unexpected but beneficial regulative side effect," illustrating once again that not all unintended consequences of genetic engineering are negative.

■ Simultaneously with the public-sector work on GR 0.5, a group of Syngenta scientists was working on a third version of the rice, a nutritionally enhanced variety based on yet another new genetic construct. The construct incorporated the daffodil *psy* gene as in the two previous versions of the rice, but it lacked *any* type of selectable marker gene, antibiotic or otherwise. The scientists produced several examples of the type in a javanica cultivar rice named Cocodrie, a high-yield, early-maturing, long-grain rice that had been developed at Louisiana State University in the 1990s. The Cocodrie specimens expressed far higher levels of beta carotene than did either of the previous versions of the rice, and for that reason this version became known as GR 1. It was soon eclipsed, however, by the fourth, final, and current version of Golden Rice, GR 2. Golden Rice 2 expresses such greatly increased levels of beta carotene, up to 37 micrograms per gram, that GR 1 was virtually wiped out of the historical record.

With four versions of Golden Rice now in existence, the challenge was to get the rice out of the lab and to the farmer. That transition would not be easy . . . or fast . . . or straightforward. In fact, it would prove to be the most difficult and problematic undertaking in the long career of Golden Rice development.

Thus far, all the different versions of the rice, and all of the experimental plants that had been produced, grown to maturity, and yielded up second and third generations of countless individual rice plants, all of them had been creatures of the laboratory or the greenhouse. They were the plant-world equivalent of lab rats.

So long as they were confined to these artificial environments and kept away from the general public, these transgenic rice plants were considered "safe" by regulators: they were hidden away and physically isolated from the outside world in the botanical version of solitary confinement. But, at least in the view of these same regulators and regulatory agencies, letting the plants out of the lab and into the real world of sunlight, fresh air, rain, and ordinary people was something of a risky proposition. Regulators viewed these genetically engineered specimens as potentially hazardous, a threat to human health and the environment.

The curious part about this, however, was that regulatory bodies had no actual *proof* that these plants were dangerous. They didn't even have any good reason to believe that the rice or the plants that produced it were *likely* to pose any such threat. What was most fundamentally behind their fears was that these organisms might have "unknown" consequences, risks, or threats. Thus it was the *lack* of knowledge of any danger rather than the *possession* of knowledge of a threat that was the chief motivating factor in the policy of regulating GMOs strictly and repressively, in a way that delayed their appearance in farmers' fields, and in the marketplace, and in the mouths of people who could benefit from their use.

It was anomalous to base regulatory policies and decisions on an absence of knowledge, on unsupported and generalized fears that genetically modified organisms just *might* be harmful but for no known or specifiable scientific reason. Fear of the unknown is a powerful force, one that can easily get out of hand and become paranoiac. Basing policy decisions on unknowns is therefore itself a potentially dangerous activity. But there were historical reasons why such a decision procedure came into existence and became widely adopted among nation-states. The practice stemmed from a UN-sponsored resolution known as the Cartagena Protocol and its central doctrine, the apparently benign but in reality quite dubious Precautionary Principle. If there was a single greatest obstacle that prevented Golden Rice from reaching those who needed it, it was this.

THE PROTOCOL

The Cartagena Protocol is its short name. But like many other treaties, international agreements, conventions, covenants, compacts, concordats, and the like, it also has a longer and more stately formal title: The Cartagena Protocol on Biosafety to the Convention on Biological Diversity. It is an international agreement, sponsored and developed by the United Nations, that aims "to ensure the safe handling, transport and use of living modified organisms (LMOs) resulting from modern biotechnology that may have adverse effects on biological diversity, taking into account also risks to human health." Superficially sensible and even innocuous, it, together with other UN instruments and institutions, would have a major, determining role in the fate of genetically engineered organisms, particularly Golden Rice.

The Cartagena Protocol is a sweeping set of guidelines, requirements, and procedures that are legally binding on the nations that are parties to the agreement, coupled with a set of mechanisms to enforce and ensure compliance with its terms. The text of the protocol consists of a preamble, followed by 40 separate articles, each of which contains one or more enumerated provisions, and ends with three annexes covering subjects not fully addressed in the articles themselves.

Although most of its provisions had been stated in draft form by the Open-Ended Ad Hoc Working Group on Biosafety, which met several times between July 1996 (in Aarhus, Denmark) and February 1999 (in Cartagena, Colombia), the finished and final text was approved and adopted only at a supplementary meeting held at Montreal, Canada, on

January 29, 2000. For that reason the treaty might more fittingly be called the "Montreal Protocol," but a prior, 1987 Montreal Protocol already existed.

The document was opened for signature by the world's governments on May 15, 2000, the first signatory being Kenya. Three years later, in May 2003, the 50th nation signed and ratified the protocol, which finally entered into force on September 11, 2003. As of 2018, the treaty has been signed and ratified by a total of 171 nations. Among the 26 nations that are *not* parties to the agreement, and therefore are not bound by its terms, 3 are notable: the United States, Canada, and the Russian Federation.

If there were a single master rule or overall governing law that formed the core of the protocol and shaped its provisions, it is the so-called Precautionary Principle (or precautionary "approach"). This is a doctrine with a history. An early form of the Precautionary Principle had been advanced at a previous UN meeting known informally as the Earth Summit but officially called the United Nations Conference on Environment and Development. Held in Rio de Janeiro in June 1992, the conference was a massive affair attended by representatives of 174 governments, 2,400 NGOs, and 17,000 members of the general public, for the purpose of defining and bringing into existence a sort of new world order.

Consistent with such grandiose aims, the Earth Summit created and promulgated a total of six new documents and legally binding international agreements. One of the latter was the Convention on Biological Diversity, which was conceived as a landmark attempt to protect the global variety of plants and animals and to promote the sustainable use of biological resources. In an effort to do so, the preamble states as a general principle that, "where there is a threat of significant reduction or loss of biological diversity, lack of full scientific certainty should not be used as a reason for postponing measures to avoid or minimize such a threat."

The principle was unusual in the first place for the considerable epistemological tension created by stating *both* that "there is a threat" *and* that "scientific certainty" regarding that threat might be lacking. If full scientific certainty that a threat exists is lacking, then how can it nevertheless be categorically asserted, as the preamble does, that "there *is* a threat"? For the sake of consistency, if certainty is lacking, then the only claim that would be epistemologically warranted is that "there *might be* a threat." But, as will become clear, exaggeration, overcaution, and pessimism (as well as, in some cases, paranoia), are hallmarks of the many efforts designed to protect the world from environmental threats, real or imaginary.

The second oddity was that the principle's focus on the *lack* of knowledge or certainty nevertheless became, upon the principle's actual implementation in practice, a justification for imposing the most stringent possible requirements on whatever was thought to pose the alleged threat. This had the consequence, in some cases, of an indefinite postponement of a product or technology causing more damage than the uncertain, and possibly nonexistent, threat may have posed.

Despite the incongruity of its central problematic claims, all United Nations member states attending the Rio Earth Summit, including even the United States, signed the Convention on Biological Diversity. The United States, however, failed to ratify the treaty, so its provisions never became binding upon this country, which, other than for the Holy See, was and remains the sole holdout.

It was another Earth Summit document, though, that contained the classic, orthodox, and most often-cited formulation of the Precautionary Principle: the Rio Declaration on Environment and Development. The Rio Declaration set forth a grand total of 27 principles that collectively had "the goal of establishing a new and equitable global partnership through the creation of new levels of cooperation among States, key sectors of societies and people." Principle 15 embodies the Precautionary Principle: "In order to protect the environment, the precautionary approach shall be widely applied by States according to their capabilities. Where there are threats of serious and irreversible damage, lack of full scientific certainty shall not be used as a reason for postponing cost-effective measures to prevent environmental degradation."

Principle 15 is not binding on the United States, nor indeed on any other country, because the declaration is not an international treaty or a legally binding instrument but merely an "agreement" that lacks the force of law. In fact, what the Rio Declaration really amounted to was an aspirational set of principles, a collection of hopes. The document was never offered to any nation for signature or even properly voted on. Rather, on June 14, 1992, the assembled national governments merely "adopted" the Rio Declaration, essentially by consensus, even though many countries had first lodged objections to or had expressed reservations about one or more of its principles.

A peculiar feature of Principle 15 was that it spoke only about preventing "environmental degradation" and said nothing about possible risks to human health or well-being. That defect was remedied eight years later, in 2000, by the Cartagena Protocol. The protocol is an attachment to the Convention on Biological Diversity that adds language pertaining to the potential threats to human health or the environment posed by the products of the genetic engineering of organisms. The protocol states its overall objective in Article 1, which reads in its entirety:

> In accordance with the precautionary approach contained in Principle 15 of the Rio Declaration on Environment and Development, the objective of this Protocol is to contribute to ensuring an adequate level of protection in the field of the safe transfer, handling and use of living modified organisms resulting from modern biotechnology that may have adverse effects on the conservation and sustainable use of biological diversity, taking also into account risks to human health, and specifically focusing on transboundary movements [i.e., national border crossings].

The Precautionary Principle itself is stated separately:

> Lack of scientific certainty due to insufficient relevant scientific information and knowledge regarding the extent of the potential adverse effects of a living modified organism on the conservation and sustainable use of biological diversity in the Party of import, taking also into account risks to human health, shall not prevent that Party from taking a decision, as appropriate, with regard to the import of the living modified organism . . . in order to avoid or minimize such potential adverse effects.

As with other versions of the principle, this formulation also allows countries to restrict, postpone, or ban a product or technology without adequate, or indeed even *any*, evidence of its likelihood of causing harm.

Whatever its virtues or defects, the Precautionary Principle (or "precautionary approach") became legally binding on all of the parties to the Cartagena Protocol. Still, exactly what the Precautionary Principle means, how it differs from the "precautionary approach," and how precaution is to be understood, interpreted, or exercised in any given case was unclear at the time and has remained so since. The principle would become fodder for endless scholarly debates conducted in journal articles, entire books, conferences, workshops, and in courts of law. It would be dismissed by some as meaningless, as paralyzing, as being too vague to offer any practical or realistic guide to action. And it would be defended and extolled by others as an eminently sound, sane, and even indispensable ground rule of risk management.

But whatever it means, and however it is to be interpreted, what has become clear over the years since it was first set forth as a regulative principle of human action is that its actual implementation has persistently erred on the side of caution, overregulation, restriction, and prohibition regarding practically every aspect of research, testing, and release into the environment of genetically engineered plants and animals. That partiality for overcaution has led in turn to a series of unintended negative consequences in the form of protracted delays, overly burdensome procedures, increased costs of compliance, and regulatory encumbrances and restrictions that in many cases are highly disproportionate to the magnitude of the potential threat, if any.

Paradoxically, the steadily mounting costs and burdens of compliance with the Precautionary Principle have themselves become "adverse effects," many of which were scientifically more certain to occur, and some of which were more harmful and serious than the theoretical, unknown, and possibly even nonexistent adverse effects that the principle had been designed to avoid.

In the specific case of Golden Rice, those delays, regulations, procedures, and costs would have tragic, permanent, and irreversible effects on human beings in the form of lives and sight lost to zealous overcautiousness.

■ In the years since it was promulgated, the Precautionary Principle has become a polarizing doctrine, causing a great divide in emotional and intellectual responses, an unbridgeable gap, a no-middle-ground abyss on the order of the opposition between believers and atheists, Republicans and Democrats, or GMO advocates and GMO opponents. Which is odd for a principle that on the surface sounds so reasonable.

At first glance, the Precautionary Principle sounds like a dressed-up version of commonplace and essentially harmless platitudes such as "look before you leap," "better safe than sorry," "an ounce of prevention . . . ," and the like. But such truisms are proverbs—clichés, really—not practical guides to human action or to intelligent public policy decisions.

And so the first order of business in assessing the validity or usefulness of the Precautionary Principle (PP) is to be clear about what it actually says and means. But that is no easy task. The principle has been advanced in several other contexts besides the Rio Earth Summit, the Cartagena Protocol, and so on. There are alternative formulations in the 1998 Wingspread Statement on the Precautionary Principle, the United Nations Framework Convention on Climate Change, the Stockholm Convention on Persistent Organic Pollutants, and in several other places as well. Indeed, a scholarly study from 2000 identified no fewer than 14 separate formulations of the Precautionary Principle in treaties, declarations, and assorted proclamations. In addition, there are "strong" versions and "weak" versions of the principle: formulations that include the idea of proportionality between the degree of risk and the magnitude of the restriction imposed and others that do not.

Common to all of its various different formulations, however, is the fact that the principle is an attempt to specify how to make decisions under conditions of uncertainty, coupled with a general assumption that under those conditions it is safer to limit a new activity, product, or technology than to permit it. This means that the principle is inherently conservative, even regressive, and acts to preserve the status quo rather than to promote change or to foster technological innovations that could end up being beneficial.

It is an understatement to say that the Precautionary Principle is controversial; in fact, it has been the subject of such an immense debate across so many years that it is impossible to do justice to the relevant issues in any sort of brief summary. The best one can do in a short space is to highlight a few of the most basic points and representative arguments pro and con.

The principle has come under sustained attack by a long series of critics. For instance, in a 1999 letter to *Nature* ("Precautionary Principle Stifles Discovery"), authors Søren Holm and John Harris, bioethicists, argued that "the PP will block the development of any technology if there is the slightest theoretical possibility of harm. So it cannot be a valid rule for rational decisions."

In 2000, François Eisinger, a professor of public health, pointed out in *Science* ("Precautionary Principle: A Self-Defeating Concept?") that "the Precautionary Principle leads to risk transfer not risk eradication. For example, if the use of crops genetically engineered to be resistant to parasitical organisms were banned, other processes such as the application of pesticides would take the place of the genetic modifications. Banning genetically engineered crops is to promote the use of chemical pesticides."

And in 2003, Harvard Law School professor Cass Sunstein wrote (in "Beyond the Precautionary Principle") that "the principle should be rejected, not because it leads in bad directions, but because it leads in no directions at all. The principle is literally paralyzing—forbidding inaction, stringent regulation, and everything in between. The reason is that in the relevant cases, every step, including inaction, creates a risk to health, the environment, or both."

But, of course, the principle has also had its defenders. Perhaps the clearest point in its favor is the incontrovertible fact that lack of sufficient caution can sometimes lead to disastrous results, and in fact has done so in the relatively recent past, as a wealth of stock examples amply attests. The poster-boy case here is the history of the drug thalidomide.

Thalidomide was first marketed in 1957 by the German pharmaceutical company Grünenthal as a sedative and sleeping pill that was also useful for the alleviation of morning sickness in women during the first trimester of pregnancy. The drug was extremely popular for a while, and at one point was being sold in more than 20 countries in Europe and Africa. In 1960, the American drug firm Richardson-Merrell submitted to the FDA a New Drug Application for the sale of thalidomide in the United States.

But by 1961 it had become clear that its use caused spontaneous abortions, stillbirths, malformed limbs, and other horrific birth defects, after which Grünenthal pulled the drug from the marketplace. And in March 1962, while the FDA was still considering it, Richardson-Merrell withdrew its New Drug Application for thalidomide. The episode became a potent argument for a more cautious approach to drug approval. If there ever were a knockdown case for the Precautionary Principle, it was thalidomide. So powerful an example is it of the need for adequate caution that it possesses an almost coercive persuasive force.

A second example illustrating essentially the same point is mad cow disease, technically known as bovine spongiform encephalopathy (BSE), in England. Called "mad cow" disease because cattle suffering from it tend to drool and to walk with an unsteady, uncoordinated gait, the first cases had appeared in England in 1986. BSE is a neurodegenerative disease with a long incubation period that ranges from three to five years or more in adult cattle, and it is invariably fatal. It is thought to be caused by cows eating meat and bone meal made from the carcasses of cattle previously afflicted with the disease.

For several years the British government consistently and adamantly claimed that the epidemic posed no danger to human health and specifically denied that BSE was transmissible to humans by eating beef. By 1990 more than 10,000 cases of mad cow disease had been reported in England, but to reinforce the view that British beef was "perfectly safe," agriculture minister John Gummer went so far as to pose with his four-year-old daughter, Cordelia, eating hamburgers. But, in 1995, the first known human casualty of what is now called "new variant" Creutzfeldt-

Jacob disease (nvCJD, also nCJD), died from eating contaminated beef. The victim was nineteen-year-old Stephen Churchill, a Royal Air Force cadet. Additional cases of nvCJD soon followed. The fact is that British beef was *not* perfectly safe, and eating the meat of diseased cattle could and did cause a similar disease in humans.

From such cases it would appear that greater precaution, and specifically greater regulation, means greater safety, and that therefore strict adherence to the Precautionary Principle is justified. But in discussions of the Precautionary Principle, there is always another side to the story. One problem here is knowing how much to generalize from these examples, which may be atypical and for that reason may lose much of their apparent force upon a deeper and more comprehensive evaluation and analysis. A second problem is that of the hidden *costs* that may be imposed by overcautiousness in cases that are less out of the ordinary. These factors serve to dilute and weaken the apparent force of the two examples cited.

The mad cow disease–Creutzfeldt-Jacob disease connection is unusual for two reasons. The first is the long incubation period of both BSE and nvCJD, which made it hard for scientists to establish a causal linkage between the two diseases. Another is that the infectious agent (thought to be a tiny protein particle, the prion), is invisible in light microscopes, contains no DNA, and is not destroyed by cooking. In consequence, it was difficult for scientists to identify the causal agent with precision.

It would be wrong, therefore, to invoke rigorous, generalized, and widespread precautionary measures on the basis of a case or cases characterized by extraordinary or unique features such as these. In instances where a given new product, procedure, or technology is substantially similar to one that is known to be safe, the need for precaution is arguably less stringent. Golden Rice, in particular, is almost 100 percent identical to common and ordinary rice except for the presence of beta carotene in the endosperm. Beta carotene is present in the rest of the rice plant, and in several fruits and vegetables, in milk and butter, and is generally recognized as safe. Overcautiousness about Golden Rice in the face of these facts is extremely hard to justify.

More generally, the safety provided by overcaution is not free and in fact often exacts substantial costs of its own. For example, a delay in approving a new drug, procedure, or technology carries its own risks and burdens that may themselves turn out to be harmful, even lethal. In a series of articles, economists Daniel Klein and Alexander Tabarrok, of George Mason University, surveyed the negative consequences of the FDA requirement that a drug be proven safe and effective prior to approval. One result is that extensive testing delays the availability of a drug that could otherwise be helping people. In the interim, some patients who would have been assisted by the drug, or had their lives saved by it, end up remaining sick, getting worse, or dying. Second, a requirement for general, large-scale, and prolonged testing programs can drive up the costs of bringing a given drug to market to the point that some potential drugs are not in the end developed at all, which in turn means that those who could have been aided by the medications are not.

In addition to those theoretical points, Klein and Tabarrok provided three separate bodies of empirical evidence to demonstrate that the costs of FDA precautionary requirements are often greater than their benefits. The first consists of the rate of new drug introductions before and after 1962, the year of the Kefauver-Harris Amendments to the Food, Drug, and Cosmetics Act of 1938, which substantially increased the scope and stringency of FDA drug approval regulations. Prior to 1962, an average of *40* new drugs were introduced per annum, with a *two*-year average time to approval; after that date the number had dropped to *16* new drugs approved per year, even though the number of new drugs proposed remained about the same as previously, with an approximate *eight*-year time to approval. "The delay and large reduction in the total number of new drugs has had terrible consequences," wrote Klein and Tabarrok. "It is difficult to estimate how many lives the post-1962 FDA controls have cost, but the number is likely to be substantial."

The second body of evidence compares drug availability, time to approval, and safety in the United States against the same set of variables in European countries. Generally, drug approvals are faster, and drugs are usually available in Europe before they are in the United States—this is the so-called drug lag. But the lag in drug approval does not result in greater safety, as judged by the rate of post-market drug safety withdrawals, which are about equal here and abroad, averaging 3–4 percent in both cases. In fact, the drug lag brings greater risk to patients for whom the delayed drug is unavailable for use.

The third body of evidence presented by Klein and Tabarrok compares on-label and off-label uses of drugs. When the FDA approves a drug, the approval is for a given, specified (on-label) use. But doctors routinely, and legally, prescribe drugs for other, off-label uses. When prescribed for an off-label use, the patient is in effect receiving a non-FDA-approved drug for a given ailment. But despite the unapproved off-label use, drugs are generally no less safe when used for off-label purposes than for on-label conditions. Lack of FDA approval for off-label use does not translate into greater risk or reduced safety.

All of the harms in these examples are a function of delays caused by overregulation of pharmaceuticals. But the general point illustrated has wider application because delays can cause harm to those who would be aided by *any* type of new product, food, device, or technology. Overcautiousness in whatever sphere of human activity, innovation, and invention imposes corresponding risks and burdens, and causes its own set of harms.

■ In addition to the empirical record of harms resulting from precautionary delays, critics claim that the Precautionary Principle suffers from a number of further defects. For example, because it focuses on risks whose precise nature, magnitude, and probability of occurring are unknown, the principle allows restrictions to be imposed without any actual evidence that harms will in fact be caused, or if there is, without any estimate of how serious they are likely to be. Further, the principle focuses on theoretical or potential risks, those that are only possible or hypothetical, while ignoring the specific and actual harms that restrictions or prohibitions are likely to produce.

The principle invokes the specter of irreversible harms, and while many harms are truly irreversible, including serious illnesses and deaths, other harms are reversible, although they might be difficult or expensive to reverse. Even when effects are in fact irreversible, that alone does not mean they are significant. The fallout of the so-called StarLink scandal is often cited as an example of an irreversible effect. StarLink was a type of genetically modified maize that in 1998 had been approved for use as animal feed in the United States. But farmers were unsuccessful in keeping StarLink corn isolated from non-GMO corn, and in the year 2000 traces of StarLink corn were detected in taco shells and corn chips. Star-Link's manufacturer, Aventis, bought up the entire harvest of its GMO maize and withdrew the product from the marketplace. Nevertheless, contaminated corn continued to turn up everywhere, prompting the aphorism "StarLink is forever."

But in fact it didn't matter in the least, either to human health or to the environment. Twenty-eight people complained of allergic reactions to the affected taco products, but an investigation by the Centers for Disease Control determined that their allergies were not to the Cry9C protein found in StarLink but rather to various allergens. Moreover, StarLink maize failed to invade cornfields and to displace other varieties, which was a common fear of critics who regarded GMOs as rapacious invasive species. Thus, although the infiltration of StarLink corn into the food chain seemed to be at least a *temporarily* irreversible effect, it was one that caused no harm, illness, or injury of any kind and had no significance other than as a public relations disaster and a financial setback for Aventis, which lost more than $100 million in the episode.

The benefits forgone by delaying a technology, by contrast, *are* irreversible, because they are in the past and cannot be recaptured. Sunk costs are lost forever. Those whose lives or sight could have been saved by earlier approvals and fewer regulatory obstacles to the development of Golden Rice, for example, are irreversibly and irrecoverably gone, and the loss of their lives and sight is significant.

The Precautionary Principle is further criticized for its tunnel vision that focuses on worst-case scenarios and ignores an entire range of other possible outcomes, including those that are beneficial rather than harmful. In 2004, after the Cartagena Protocol came into force, the United Nations Environment Programme (UNEP) was spending more than

$100 million on training personnel in developing countries to conduct risk assessments in relation to "transboundary movement of living modified organisms," that is, on the risks attached to GMOs crossing national borders. At that time, Adrian Dubock, a founding member of the Golden Rice Humanitarian Board, asked the leader of the UNEP Biosafety Unit how much was being spent on *benefit* assessment of GMO crops. The answer was, nothing. That is tunnel vision.

The tendency of the Precautionary Principle's advocates to invoke worst-case scenarios is vividly illustrated by a recent analysis, "The Precautionary Principle (with Application to the Genetic Modification of Organisms)" (2014), by Nicholas Taleb and others. Taleb, a professor of risk engineering at NYU, is the author of *The Black Swan* (2007), about "the impact of the highly improbable," and is the cofounder and codirector of the Extreme Risk Initiative. In a working paper, Taleb and four other scholars confined the Precautionary Principle to a narrow subset of events called " 'ruin' problems, in which a system is at risk of total failure . . . [and] outcomes may have infinite costs."

Genetically engineered organisms, these scholars claimed, fall into the category of phenomena that could have infinite costs and therefore should be subject to severe limitations. Their "potential harm is systemic (rather than localized) and the consequences can involve total irreversible ruin, such as the extinction of human beings or all life on the planet." This is a worst-case scenario par excellence.

But if extraordinary claims require extraordinary evidence, then Taleb and his colleagues failed to provide it. They distinguished between localized, non-spreading harms (for example, nuclear core meltdowns) and nonlocalized, systemic, propagating impacts (like kudzu) and then went on to place GMOs in the latter category. But the authors did little more than *assume* that the systemic, propagating impacts of GMOs will be harmful. They spoke, for example, of "the largely unknown risks associated with the introduction of GMOs." The claim of "unknown risks," as we have seen, is one of the most common criticisms leveled against GMOs by their opponents. But to say that the consequences of their use are *unknown* is at the same time one of the weakest arguments against

GMOs in terms of evidentiary value. If a risk is truly "unknown" in nature, size, duration, seriousness, and scope, then how is it possible to judge it one way or the other? Indeed, if the effects of GMOs are "largely unknown," then how can they meaningfully be classified as risks as opposed to benefits or as having consequences that are neutral, neither good nor bad? It is like assessing the value of an intrinsic nothing.

Further, Taleb and the other authors tried to make the worst of any uncertainty. "An increase in uncertainty leads to an increase in the probability of ruin," they wrote. But this, too, is epistemologically untenable: an increase in uncertainty is just that and leads to a reduced degree of probability of *any* given outcome, positive or negative.

The authors contended, "A lack of observations of explicit harm does not show absence of hidden risks." Nor does it show the *presence* of hidden risks or even suggest the *potential* of risk.

The authors considered the case of Golden Rice, but they presented no more than an ad hominem argument against it: "Given the promotion of 'golden rice' by the agribusiness that also promote biofuels, their interest in humanitarian impacts versus profits gained through wider acceptance of GMO technology can be legitimately questioned." This is guilt by association, not an argument from evidence.

The authors, in short, assumed the very worst at every turn, defended their position with a series of non sequiturs, and in the process failed to make an effective case either against GMOs or for the Precautionary Principle.

■ At this point, one important matter is still left dangling: thalidomide. Isn't thalidomide a genuine example of a worst-case scenario actually coming to pass in Europe through lack of adequate caution? Isn't its prevention in the United States due precisely to sufficient caution in fact being exercised, with approval of thalidomide denied by the FDA?

There is no question that the use of greater caution in Europe would have prevented the wave of birth defects, stillbirths, and deaths caused by thalidomide. But the situation is more complicated when it comes to the United States. It is popularly thought that it was cautiousness on the

part of Frances Kelsey, the US FDA inspector assigned to review thalidomide for possible release in the United States, that led her to deny
Richardson-Merrell's application for the drug. But, in fact, neither she
nor the FDA itself ever formally denied approval of thalidomide. Rather,
it was the Richardson-Merrell company that unilaterally withdrew its
application after proof of the harms caused by thalidomide had accumulated. This is acknowledged by Frances Kelsey herself, who in her "Autobiographical Reflections" (2014) said, "On 8 March 1962 the formal withdrawal of the application was submitted. There is nothing that would
lead me to think we had requested the withdrawal. I think Merrell withdrew it of their own accord when they were finally convinced that there
really was a problem related to the drug."

It is important to recognize that the company's decision to withdraw
its application was a product of increasing empirical evidence that the
drug was unsafe for pregnant women, as opposed to a generalized precautionary effort to avoid possible, theoretical, or unknown harms or
risks. In any case, one cannot take the single example of thalidomide
and infer from it the more sweeping conclusions either that FDA approval equals safety or that thalidomide is categorically "unsafe" and
should be banned for all people and for all time.

On a worldwide basis, an estimated 10,000 infants died or suffered
birth defects due to thalidomide during the period that the drug was on
the market, although this number is not known with precision. By contrast, a meta-analysis published in 1998 showed that in a single year, 1994,
at least 106,000 people died from adverse reactions to FDA-approved
"safe and effective" drugs in the United States, making such reactions the
sixth leading cause of death in this country. Further, that number pertained only to hospitalized patients and did not include those who died
of adverse reactions at home or elsewhere. A later (2000) study by other
authors added the information that 350,000 significant adverse drug reactions occur in US nursing homes each year.

From this data it follows that FDA approval of a drug as "safe" does
not really mean that it is harmless to all those who take it. Perfect safety
is a will-o'-the-wisp and an illusion.

Nevertheless, one could perhaps breathe a huge sigh of relief in the knowledge that at least thalidomide was never approved for sale in the United States.

Except that it was.

Thalidomide is not banned in the United States; it is currently marketed here for both on- and off-label uses. There are two reasons for this. One is that thalidomide poses health risks chiefly to pregnant women and their unborn fetuses. For them it is extremely unsafe. But that leaves all the rest of the human population, an entire and very large group of men and nonpregnant women for some of whom the drug provides actual benefits.

Drug effectiveness is a two-edged sword, with serious side effects for some people coexisting with efficacy against certain illnesses in others. It turns out that, for all the damage it does to unborn fetuses, thalidomide is nevertheless useful for many illnesses, ranging from rare diseases to more common conditions. For example, it has powerful anti-inflammatory properties that make it effective against a painful inflammatory complication (*Erythema nodosum leprosum*) associated with leprosy. Such an inflammation affects about 100–200 people in the United States annually, and for them the drug is not only safe but demonstrably beneficial. For that reason, the FDA approved the sale of thalidomide for the treatment of that condition in 1998, while at the same time imposing severe restrictions on its distribution to both men and women.

Further research showed that thalidomide is also effective in treating certain cancers such as multiple myeloma, a cancer of the bone marrow, a use for which the FDA approved thalidomide in 2006.

Physicians also prescribe thalidomide for off-label uses, as it was found to augment immune responses in such a way that makes it helpful in treating throat and oral ulcers in patients with HIV. It also helps alleviate the substantial weight loss suffered by AIDS victims.

Today, thalidomide is sold by the drug manufacturer Celgene under the trade name Thalomid and by other companies under other names. Thalidomide is currently available, at great expense, at ordinary local pharmacies such as CVS, Walgreens, Rite Aid, Costco, and Walmart.

Celgene's white 50-milligram Thalomid capsules are imprinted with an image of a pregnant woman over which is superimposed the international prohibition sign (circle backslash symbol).

There are several morals to the story of thalidomide and the wider class of drugs to which people may have adverse reactions. One is that FDA approval does not guarantee "safety." Safety is not an all-or-nothing category, and what is safe or even healthful for some may be extremely dangerous or fatal to others. Another is that caution does not eliminate danger or illness, for, during the time that thalidomide was effectively banned in the United States, all those patients who could have benefitted from its use for a variety of medical conditions were unable to.

But the most important lesson of all is that precaution must itself be used cautiously, because overcautiousness can and does exact a price of its own.

■ The Cartagena Protocol was formally adopted on January 29, 2000. A little more than a year later, on March 12, 2001, but well before the protocol would come into force on September 11, 2003, the European Parliament issued Directive 2001/18/EC "on the deliberate release into the environment of genetically modified organisms." This is a landmark piece of legislation that is legally binding on all member states of the European Union. The directive outlines an encompassing set of regulations and procedures for GMO containment measures, risk assessment, labeling, packaging, and traceability, as well as penalties for noncompliance. The document, which in English runs to 25 pages set in very small type, includes, as provision (8), the statement that "the precautionary principle has been taken into account in the drafting of this Directive and must be taken into account when implementing it." Further, Article 32 of the directive, "Implementation of the Cartagena Protocol on Biosafety," mandates the use of "appropriate measures to implement the procedures laid down in the Cartagena Protocol."

The directive would go on to control virtually all significant research, development, and testing of genetically modified organisms throughout the member states of the European Union. Switzerland, not being a member state, passed a similar law in 2003, the Federal Act on Non-human Gene Technology, with its own set of restrictive provisions on the use and handling of GMOs.

A particularly onerous requirement of Directive 2001/18/EC is the enormous set of conditions that have to be fulfilled for a developer to gain permission to release a genetically modified organism into the environment, as in a field trial. Anyone wishing to conduct a field trial of a given GMO has to compile a technical dossier that provides a virtually complete set of data about the plant, how it has been modified, and all associated personnel and infrastructure, including, to start with:

   (i) general information including information on personnel and training,
  (ii) information relating to the GMO(s),
 (iii) information relating to the conditions of release and the potential receiving environment,
  (iv) information on the interactions between the GMO(s) and the environment,
   (v) a plan for monitoring in accordance with the relevant parts of Annex III in order to identify effects of the GMO(s) on human health or the environment,
  (vi) information on control, remediation methods, waste treatment and emergency response plans,
 (vii) a summary of the dossier.

Annex III (mentioned in (v), above), lists the detailed information required concerning the genetic modification itself and the resulting genetically modified plant:

C. INFORMATION RELATING TO THE GENETIC MODIFICATION
   1. Description of the methods used for the genetic modification.
   2. Nature and source of the vector used.
   3. Size, source (name) of donor organism(s) and intended function of each constituent fragment of the region intended for insertion.

D. INFORMATION RELATING TO THE GENETICALLY MODIFIED PLANT

1. Description of the trait(s) and characteristics which have been introduced or modified.
2. Information on the sequences actually inserted/deleted:
    (a) size and structure of the insert and methods used for its characterisation, including information on any parts of the vector introduced in the GMHP [genetically modified higher plant] or any carrier or foreign DNA remaining in the GMHP;
    (b) in case of deletion, size and function of the deleted region(s);
    (c) copy number of the insert;
    (d) location(s) of the insert(s) in the plant cells (integrated in the chromosome, chloroplasts, mitochondria, or maintained in a non-integrated form), and methods for its determination.
3. Information on the expression of the insert:
    (a) information on the developmental expression of the insert during the lifecycle of the plant and methods used for its characterisation;
    (b) parts of the plant where the insert is expressed (for example roots, stem, pollen, etc.).
4. Information on how the genetically modified plant differs from the recipient plant in:
    (a) mode(s) and/or rate of reproduction;
    (b) dissemination;
    (c) survivability.
5. Genetic stability of the insert and phenotypic stability of the GMHP.
6. Any change to the ability of the GMHP to transfer genetic material to other organisms.
7. Information on any toxic, allergenic or other harmful effects on human health arising from the genetic modification.
8. Information on the safety of the GMHP to animal health, particularly regarding any toxic, allergenic or other harmful

effects arising from the genetic modification, where the GMHP is intended to be used in animal feedstuffs.

9. Mechanism of interaction between the genetically modified plant and target organisms (if applicable).
10. Potential changes in the interactions of the GMHP with non-target organisms resulting from the genetic modification.
11. Potential interactions with the abiotic environment.
12. Description of detection and identification techniques for the genetically modified plant.

On a global level, regulations pertaining to GMOs are so diverse and fragmented that it is virtually impossible to make blanket generalizations about them, and it is equally difficult to characterize them individually; in addition, more than 100 nations have no such regulatory systems to begin with. A recent (2015) study of "patterns and determinants of GMO regulations" ranks the restrictiveness of the GMO regulatory systems of 60 countries. Hong Kong and Bangladesh were among the least restrictive, while several countries within the European Union, nations with some of the world's most advanced technologies, had some of the most stringent regulations of all.

It is perverse that application of the Precautionary Principle and the sets of directives and requirements created in its wake force GMO developers to amass extremely detailed and voluminous stores of knowledge about the plant they present to regulators for approval, but that at the same time, such approval can be denied not on the basis of proof or knowledge of likely harm posed by the plant in question, but merely on the basis of "unknown risks," that is, by the *lack* of knowledge, evidence, or data.

As it is, the European directive, along with the corresponding regulatory systems in other countries, has two main consequences for GMOs. It drives up the cost of preparing any GMO for field trials, and it increases the time between the first genetic modification to an organism and the time at which the first field trials of the original plant or a successor can be conducted. This is unfortunate because actual, outdoor, real-world field trials are vital to the process of gaining knowledge about and optimizing the performance of a plant: they are necessary to know how any given genetically modified plant would actually perform in farmers' fields.

For Golden Rice, the final effect and ultimate cost of the Cartagena Protocol's Precautionary Principle is a series of delays that would turn the rice into a kind of reverse mirror image of thalidomide. If the thalidomide tragedy and its trail of malformed fetuses were the wages of inadequate cautiousness, it was the reverse, extreme overcautiousness, that led to equally tragic consequences in the form of the preventable blindnesses and the needless deaths suffered by those who could have benefitted from Golden Rice during the years of its protracted regulation-imposed delays.

WHAT IS A GMO?

Billions of words have been written about genetically modified foods and other organisms, and countless arguments have been made about them, for and against. But hidden beneath that vast ocean of rhetoric lurks the tacit assumption that we know exactly what a genetically modified organism is to begin with. In actual fact, however, this is not so clear. "Genetically modified organism," after all, is not a mathematical concept with a definition so precise as to admit of no ambiguity, nor is there a decision procedure that allows us to establish objectively and definitively whether a proposed example falls inside or outside the boundaries of the term. Rather, "genetically modified organism" is a phrase in common usage, and many who use the term have no clear idea of what it means, or exactly which foods or organisms qualify for the distinction, or why or why not.*

This is a crucial issue, because if we can't clearly say what a GMO is, then how can we be either in favor of, opposed to, or even neutral about the creation and use of such organisms in an intelligent, principled, and rational fashion? Obviously, we can't. And so the first question to answer in any attempt to evaluate the safety or riskiness of genetically modified organisms, and Golden Rice in particular, is the most fundamental one of all: What is a GMO?

It is important to recognize at the outset that there is a trivial sense in which *all* organisms have been genetically modified, in the first case through the ordinary processes of evolution by natural selection. Such

---

* For that matter, many people are not very well acquainted with the concept of genes, either. In 2004, researchers at Rutgers University conducted a national study which showed that 60 percent of Americans polled did not know that ordinary tomatoes contained genes.

modifications occur through spontaneous mutations, DNA replication accidents or errors, or energetic particles, such as cosmic rays or x-rays, striking a cell's genome and minutely altering, scrambling, or damaging some of its contents. Largely, the effects of such random mutations are detrimental to the organism, just as random changes to computer software or to the components of a precision machine, such as an automobile engine, don't usually lead to improved performance. But, in rare instances, random genetic mutations do act to enhance a desirable trait or suppress an undesirable one. And, in this way, such spontaneous changes produce genetic diversity and function as key drivers of evolution. In fact, random mutations are in large part responsible for the diversity and variety that we see across the universe of living things.

There is also a second trivial sense in which many organisms, including most if not all of the foods we eat, have been further genetically modified by human intervention in the form of artificial selection, that is, by traditional breeders preferentially fostering the development of certain desired characteristics in crops, flowers, farm animals, dogs, cats, or whatever else. Indeed, some forms of selection occur even without human intervention. In their book *Dealing with Genes*, the molecular biologists Paul Berg and Maxine Singer pointed out that "it is not only humans that affect the genetic traits of other species. Organisms of all types influence the genomes of other species, including those of humans: subtly (for example, by competing for certain food supplies and thereby applying selective pressure for traits that permit the use of alternate nutrients) and boldly (by killing off large populations that lack the genetic makeup to protect themselves)."

Less dramatically, the intentional crossbreeding of plant types routinely gives rise to "new and improved" varieties of fruits and vegetables. Since the changes that take place in crossbreeding occur at the genetic level, there is no getting around the fact that the resulting food products are genetically modified organisms. For this reason, many people who are in favor of GMOs, including most scientists, have argued that practically all the foods we eat, plant and animal, have already been genetically modified, and so all of them are GMOs, thus entirely wiping out the distinction between GMO and non-GMO foods.

That argument, however, overlooks the fact that, even in common usage, "GMO" stands for something distinct from the products of traditional breeding. The crossbreeding of plants usually takes place within members of the same species and results in a multiplicity of subspecies and many different individual varieties. For example, there are thousands of genetically different rice varieties, all of which are members of the same species, *Oryza sativa*, and there are some 20,000 different varieties of corn, all of which are members of the same species, *Zea mays*, and so on.

And so in many people's minds, what distinguishes a true GMO from a conventionally bred plant variety is that the GMO contains genetic material from a different species altogether. Indeed, GMOs are called "transgenic" (literally, "across genes") precisely because they contain genes from an entirely unrelated and "foreign" organism.

Still, that alone is not sufficient to make an organism a GMO for the simple reason that even traditional breeding can cross species boundaries and introduce genetic material from one species into the DNA of another, distinct species. Animals known as *crossbreeds* contain genes from different species. Mules, for example, are a hybrid cross between a female horse and a male donkey. And, in the plant world, a surprising number of the foods that we routinely eat as dietary staples are hybrid crosses, including bread wheat (a hybrid of three wild grasses), maize, peanuts, chestnuts, eggplant, sugarcane, radishes, tomatoes, grapefruit, lemons, limes, coconuts, and coffee.

Some of these crossings occurred naturally, without human intervention, while others are products of intentional or unintentional human agency. Most hybrids result from crossings between members of two different species, but some hybrids are between organisms that are not only from different species but also from different genera. Using a laboratory technique known as chemical mutagenesis, wheat (genus *Triticum*) and rye (*Secale cereale*), were crossed to produce a hybrid grain called triticale. This is certainly genetic modification, but the result is not regarded, classified, labeled, or regulated as a GMO. Indeed, flour made from a triticale grown by Azure Farm of Dufur, Oregon, is sold by the online firm Organic Kingdom as a natural and wholesome health food, even as an "organic" food, despite the claim on the Organic Kingdom web page that "organic foods are not genetically modified."

The ancient art of grafting is yet another way of producing plants and trees that combine genetic material from two different species, thereby giving rise to new genetically modified organisms. But grafting is a story unto itself. The practice is referred to in the Bible and is mentioned in ancient Greek and Chinese writings, showing that grafting was known and used in Europe, the Middle East, and Asia by as early as the fifth century BCE. Surprisingly, because of what it can, and indeed has, accomplished, grafting is so simple a process that it even occurs naturally, as when two plants touch and rub against each other and their branches or roots fuse without human aid.

Grafting can be done manually by inserting a shoot of one plant (called the scion) into the root of a different plant (the rootstock), giving birth to a new plant that is genetically different from either parent plant alone. Grafting can occur within a species or between two different species. *Three* different species—the potato, tomato, and eggplant—that are nevertheless members of the same family (the nightshade family, or *Solanaceae*), can be grafted one to another, producing fruit that's a mixture of both plants. Thus, tomatoes have been grafted with eggplants, resulting in indigo tomatoes. But the grafting of tomato and potato has yielded an even stranger creation: a unique combination plant that grows both fruits simultaneously, the tomatoes aboveground, the potatoes below. The result is the so-called pomato, the generic name for the Tomtato, which has been trademarked and is sold by the British commercial grower Thompson and Morgan.

Anyone who is aghast at the thought that modern biotechnology can introduce a flounder gene into a tomato ought to be equally horrified that the ancient biotechnology of grafting can introduce potato genes into a tomato plant, with the result that this Frankenplant can simultaneously grow the red fruit *en plein air* and the white tuber underground.

Ironically, there is a far greater difference between a Tomtato plant and either a potato or tomato than there is between Golden Rice and plain, regular white rice. But while Tomtato seeds are sold commercially and can be planted and grown in anyone's garden, Golden Rice is regarded as such an experimental, potentially risky crop that it has been confined to airtight labs or greenhouses for large stretches of its lifetime as a plant.

And although the freakish Tomtato is unquestionably a genetically modified organism, the Thompson and Morgan horticultural firm nevertheless advertises its plant as "all natural—no GM." Which once again raises the question of whether we can really say on any rational basis what a GMO actually is.

■ Crossing and grafting are two distinct ways by which humans have engaged in "genetic engineering" and have been creating genetically modified organisms long before anyone in the world even knew about genes. But now that we do know about genes, perhaps we can assert that a GMO is an organism whose genes have been modified by the instruments and techniques of modern biotechnology. Indeed, when the European Parliament issued its Directive 2001/18/EC, laying down the rules for handling GMOs, the definition it gave of the term embodied precisely that point. The directive said that the phrase "'genetically modified organism (GMO)' means an organism, with the exception of human beings, in which the genetic material has been altered in a way that does not occur naturally by mating and/or natural recombination."

A GMO, in other words, is an "unnatural" organism, a product of the laboratory, not of the greenhouse, garden, farm, or barnyard.

But consider the case of one of the most popular varieties of grapefruit, sold all over the world, even inside the notoriously GMO-averse European Union: the Rio Red, a deep red, seedless, sweet citrus fruit. It is nowhere regarded as a GMO, nor is it labeled, regulated, or sold as such. Nevertheless, Rio Red grapefruit was not the "natural" product of the greenhouse, the garden, or the farm. Instead, it was produced in a laboratory, specifically the Brookhaven National Laboratory, in Upton, Long Island, by bombardment with repeated doses of thermal neutron radiation.

Grapefruit is a fruit with a long and storied past. Unlike many of its citrus brethren, grapefruit is not native to Southeast Asia. It comes from the Caribbean island of Barbados, where it was discovered sometime in the 1700s. It originated as a natural cross between the sweet orange and the pomelo (a large, bitter pulp fruit), both of which had themselves been introduced from Asia in the seventeenth century. It was at first called "the forbidden fruit," which is a literal translation of its scientific name, *Citrus + paradisi.*[*] From its very beginning, then, the grapefruit (so named because it appears on trees in grapelike clusters), was already a genetically modified plant, inasmuch as it was a hybrid mix of two different species.

Grapefruit was brought to the United States in 1823, arriving at the port of Safety Harbor, Florida, a town just east of present-day Clearwater, and it soon became widely cultivated in the state. In the late nineteenth century, the Kimball C. Atwood Grapefruit Grove, near Ellenton, Florida, was the country's largest producer of grapefruit and was at that time shipping 80,000 boxes per year.

All of that output consisted of "white" grapefruit, characterized by a yellow outer skin and an inner pulp divided into pale yellow sections. But in 1906, grove supervisor M. B. Foster discovered among his trees a grapefruit whose inner pulp was not yellow but pink. He identified this variant as a "sport," the horticulturist's term for a spontaneous mutation. This new sport variety was now a further genetic modification of an already genetically modified hybrid. Citrus growers called this new mutant variety "nature's million dollar mistake," trading on the idea that the results of spontaneous mutations are "mistakes of nature," which remains a popular viewpoint even today. This fortunate "mistake" then began to be commercially grown in Florida, Texas, Arizona, and even India.

Then in 1929, at about the time of the stock market crash, a Texas pink grapefruit tree produced a second sport, or mutation, whose flesh was not pink but red. Texas growers now bred into existence several

* The "+" identifies the plant as a hybrid cross.

different varieties of red grapefruit, but one of them, the Ruby Red, became the most popular of them all, and it was in fact the first grapefruit to be granted a US patent. Nevertheless, the Ruby Red had a serious drawback: its color became gradually weaker during the course of the growing season. Early-harvest specimens were bright red, but by the end of the season the late fruit's color faded to pink.

At Texas A&I University (now Texas A&M), Richard Hensz, director of the university's Citrus Center, wondered whether he could force a mutation that would make the color a deeper red that persists across the entire growing season. Natural mutations, he knew, are produced by cosmic rays, x-rays, or other such influences, but they occur only rarely. Why not speed things up by duplicating and enhancing the process artificially? And so, in 1963, in an attempt to create what today we would call a "designer grapefruit," Hensz sent a few thousand Ruby Red buds to the Brookhaven National Laboratory to have them subjected to ionizing radiation in the form of thermal neutron bombardment.

The idea of what today is called "radiation breeding" was not new even then. There were two pioneers of the art: one who irradiated fruit flies, the other, plants. In the early 1920s, Herman J. Muller, of the University of Texas, thought to increase the mutation rates of fruit flies by exposing them to x-rays and radium rays. In a 1927 paper in *Science*, "Artificial Transmutation of the Gene," he described treating fruit flies "with relatively high doses of X-rays," which, he said, "induces the occurrence of true 'gene mutations' in a high proportion of the treated germ cells." The procedure, in other words, introduced a true genetic modification of the organism by an artificial laboratory technique.

Still, there was no consistent outcome of these exposures: some of the flies survived the experience and even produced live and viable offspring. Others died, while another subset of the irradiated flies showed visible mutations. "Some of these involved morphological effects of a sort not exactly like any seen previously," Muller reported. Long afterward, in 1946, Muller received a Nobel Prize for this work.

The pioneer of induced mutations in plants was Lewis John Stadler, of the Department of Field Crops at the University of Missouri. Starting in

1928, while investigating the nature and structure of plant genes, Stadler had the idea of observing the effects of ionizing radiation on the genetics of maize and barley. He subjected barley seeds to 12 separate x-ray exposures spaced at one-hour intervals. He also experimented with radium treatments at various levels and lengths of exposure. Upon germination, the irradiated seeds produced mutant plant types exhibiting a range of effects—mainly color differences. The plants were variously white, yellow, and even striped, but in most cases they died within two or three weeks.

At Brookhaven, Richard Hensz got better results, although it took him several years to develop the ideal red grapefruit, which he did by, of all things . . . *grafting*. He inserted the irradiated Ruby Red buds into a series of existing grapefruit tree rootstocks, repeating the process across several generations sequentially and over many years, until in 1976 he finally grew a fruit that he said was "five times redder than the Ruby." Since Texas A&I was in the Rio Grande Valley, he named his neutron-bombarded, grafted, mutant creation the Rio Red.

The Rio Red was at that point a genetically modified mutant fruit plant *five times over*. To wit, it was (1) the manually grafted product of (2) an artificially mutated, irradiated variant of a fruit (the Ruby Red) that was itself (3) the spontaneous mutation (or sport) of a pink grapefruit, which was in turn (4) an earlier spontaneous mutation (or sport) of (5) a white grapefruit that had originated as a natural hybrid cross back in Barbados in the 1700s. You can hardly get more genetically modified than that. But, as repeatedly modified as the Rio Red grapefruit and its series of parent organisms had been over the centuries, it is nevertheless not considered a GMO, nor is it regulated, labeled, or marketed as such.

By any rational standard, this is odd. Exposing grapefruit buds to artificial nuclear radiation means modifying its genome at random locations, introducing changes with unknown consequences and possible risks. But the arguments that GMO opponents routinely make concerning the "unknown risks" and possible dire consequences of GMOs, or at least what they *consider* to be GMOs, somehow fail to apply to genetically modified organisms created by radiation breeding, chemical mutagenesis, manual or natural grafting, and other forms of genetic modification.

A double standard is clearly at work here, one that deems GMOs, however defined, as potentially unsafe while giving a free pass to those genetically modified organisms that are nevertheless not classified as GMOs, no matter how much their genomes have been altered and rearranged by artificial or natural means, or both.

■ These days, radiation breeding and the thousands of different genetically modified (but "non-GMO") food plants that this technique has produced are a big business worldwide. A database of plants produced by radiation breeding is housed at the headquarters of the International Atomic Energy Agency (IAEA), in Vienna, Austria, as a cooperative venture with the UN Food and Agriculture Organization (FAO). The two agencies maintain the Joint FAO/IAEA Mutant Variety Database, which is publicly accessible at mvd.iaea.org. The site's welcome page reads, in part:

> The application of mutation techniques has generated a vast amount of genetic variability and is playing a significant role in plant breeding and genetics and advanced genomics studies. The widespread use of mutation techniques in plant breeding programmes throughout the world has generated thousands of novel crop varieties in hundreds of crop species, and billions of dollars in additional revenue.
>
> The FAO/IAEA Mutant Variety Database or MVD collects information on plant mutant varieties (cultivars) released officially or commercially worldwide. Data on the mutagen and dose used, the characters improved, and agronomic data if available are among the information provided. The purpose of the database is to demonstrate the significance of mutation breeding as an efficient tool for preserving and enhancing global food security, to serve as a platform for breeders to showcase their varieties to a global audience, and to stimulate germplasm transfer for cultivation, breeding or genomics studies.

Many of the plant foods that we eat every day have been genetically modified by these nuclear techniques, a process that typically uses cobalt-

60 as the source of gamma rays to artificially induce genetic mutations. The facilities in which these radiation experiments are conducted exhibit all the trappings of a nuclear research center: the buildings are replete with complex and expensive nuclear radiation machinery and equipment, the systems are surrounded by earthen or lead shielding, their temperatures are maintained by elaborate cooling systems, and so on.

The crop plants that have been produced by means of radiation breeding include some of the most prosaic and commonplace foods we eat: varieties of rice, wheat, barley, pears, peppermint, peanuts, bananas— and, of course, grapefruit. (At the Mutant Variety Database, the Rio Red variant is listed as accession number 282.) But, despite their provenance in a firestorm of nuclear bombardment, neither the resulting plants nor the foods derived from them are radioactive, harmful, or unsafe to eat.

There are yet additional methods of artificially inducing genetic modifications in plants: chemical mutagenesis, for example, which as we have seen had been used to produce the hybrid cross triticale grain. This method, in which plants or seeds are treated with synthetic chemicals, principally ethyl methanesulfonate, has several advantages over radiation breeding. For one thing, it does not require extensive nuclear apparatus, shielding, or other radiological protection methods or devices. For another, the technique is less harmful to the organism, yielding fewer damaged, aberrant, or dead plants. Finally, the method is capable of producing large mutant populations in a limited space.

At the other extreme of equipment and expense lies space breeding, a technique in which seeds are sent into space to be exposed to greater amounts of cosmic rays than would reach them on the earth's surface. Since 1987, China has used this method to produce more than 60 mutant varieties of rice, wheat, vegetables, fruits, and flowers, including a mutant lotus called the Outer Space Sun.

The development of CRISPR technology provided yet another method of artificially modifying genes.* CRISPR is a way of editing DNA molecules that typically works by rearranging the organism's own genes or by

* The acronym stands for "clustered regularly interspaced short palindromic repeats."

deleting selected base-pairs from it, rather than by introducing any new or "foreign" DNA into genes. One might regard it as a minimally invasive method of genetic modification. An application of CRISPR technology in 2015 involved altering the genes of white button mushrooms so that they would not be subject to the browning that normally occurs during storage or from bruising. A plant pathologist at Pennsylvania State University, Yinong Yang, identified the proteins that caused browning, as well as the genes that coded for those proteins. Then, by using the CRISPR gene-editing technique, he simply made small deletions (between 1 and 14 base-pairs), in the gene responsible for producing the brownish pigment, but added no foreign genes. He created, in the process, a "transgene-free, anti-browning white button mushroom."

Inasmuch as the US Department of Agriculture regarded certain genetically engineered crop plants as subject to regulation, Yang wanted the USDA itself to verify that, since his gene-edited mushroom contained no *added* genes but only *deleted* gene sequences, it would not be subject to regulation. So in October 2015 he wrote a letter of inquiry saying, "Before proceeding with further product development, I would like to seek confirmation from APHIS [Animal and Plant Health Inspection Service] that the anti-browning mushroom, which has small deletions but no integration of plant pest elements or foreign DNA sequences, is not considered a regulated article under current regulations."

Six months later, Yang got an answer. Since the anti-browning mushrooms "do not contain any introduced genetic material," nor are they plant pests or likely to "increase the weediness of white button mushrooms," APHIS did not consider the mushroom to be a regulated article.

This genetically altered mushroom is just one more in a long line of genetically modified organisms that escape regulation because they are not officially classified as GMOs. (Neither Yang's letter nor the USDA's reply explicitly raised the issue of whether the anti-browning mushroom was technically a GMO—yet a further indication, perhaps, of the difficulty of defining the term.)

And so, in summary, this is the situation: plants whose genes have been modified by (1) nature, (2) traditional crossbreeding, (3) grafting,

(4) radiation breeding, (5) chemical mutagenesis, (6) space breeding, or (7) CRISPR are not regarded, classified, labeled, or regulated as GMOs, irrespective of how much their genes have been scrambled, reorganized, or otherwise modified by any or all of those methods.*

In that case, however, what *is* regarded as a GMO?

■ The only genetic manipulation technique that *is* conventionally regarded as producing GMOs is the gene modification method variously called recombinant DNA (rDNA) technology, transgene insertion, gene splicing, or genetic engineering proper. When the European Union issued its Directive 2001/18/EC, it described the method as encompassing "recombinant nucleic acid techniques involving the formation of new combinations of genetic material by the insertion of nucleic acid molecules produced by whatever means outside an organism, into any virus, bacterial plasmid or other vector system and their incorporation into a host organism in which they do not naturally occur but in which they are capable of continued propagation."

This means that an organism is to be regarded as a GMO not on the basis of whether its genes have been *modified*—without exception, *all* the other methods also modify genes—but rather on the basis of the specific *process* used to make the modification. The only logical basis for this way of classifying organisms is that the genetic sequences inserted by recombinant methods could contain a series of diverse genes that are highly unlikely to be a product of any of the other methods. For example, the genetic construct that Potrykus, Beyer, and company inserted into the rice genome contained genes from both a flower and a bacterium, plus a promoter sequence from a cauliflower mosaic virus. But each of those genes was there for a narrow and specific purpose: the daffodil *psy* gene coded for phytoene synthase, an intermediate precursor of beta carotene. Its purpose was not to turn rice into daffodils but to

* This list does not exhaust the roster of unconventional genetic modification techniques, which also include somaclonal variation breeding, cell selection breeding, and other methods, none of which is normally regarded as producing GMOs as commonly understood.

help express beta carotene in the rice endosperm—a harmless and indeed beneficial outcome.

Ironically, despite the heterogeneous nature of many inserted genetic constructs, there is nevertheless a sense in which recombinant DNA techniques are *more* "natural" and less intrusive than the other genetic modification techniques that we have considered. For the most part, the mechanisms and agents of rDNA technology are entities or structures that already occur in natural and normal living systems: genes or parts thereof, plasmids, viral or bacterial vectors, and so on. The agents of chemical or radiation mutagenesis, by contrast, are not parts of living systems, but instead are inanimate, inert, and often destructive chemical substances or radioactive particles, some of which originate outside the earth's atmosphere, and indeed outside the solar system.

Still, recombinant technology is an extremely powerful genetic technique, because of its ability to insert into a genome one or more foreign genes from different organisms. Scientists can take genes from virtually any organism, animal, or plant and put them into any other organism, animal, or plant. Indeed, almost from the beginning, recombinant molecular methods were used to produce genetic oddities, some of which have practical value, along with a small number that don't. Some of the latter are exotic organisms created basically for their curiosity or stunt value.

Of those that have practical value, some are *chimeras*, organisms that, mimicking their namesakes from Greek mythology (fire-breathing monsters with a lion's head, a goat's body, and the tail of a serpent), contain genetic sequences from two or more different types of living entities, the elements of which are then combined in the laboratory. In 1986, for example, a group of researchers from the University of California, San Diego, genetically altered a variety of tobacco plant to glow in the dark. They performed this feat by taking the gene responsible for a firefly's light (the gene coded for the enzyme luciferase, which is responsible for bioluminescence) and adding it to the tobacco plant's genome by means of a plasmid. The result was a unique combination tobacco-firefly chimera, whose leaves indeed emit a faint glow visible to a dark-adapted eye—at least to an eye that is not vitamin A deficient.

Despite appearances, this was not done for amusement; rather, the point of the exercise was to create a reporting mechanism, or genetic marker system, a diagnostic tool that visually indicated whether, where, and when the fusion of a new gene into a preexisting genome has been successfully accomplished. (Much later, in 2014, a group of science-fiction-obsessed futurists would propose amplifying the process to the point where plants would be able to light up rooms, entire houses, and even city streets.)

Members of the same group that produced the glowing tobacco plant also managed to put the luciferase gene into live monkey cells in lab tissue cultures, for the purpose of demonstrating that the genetic marker technique also worked in mammalian cells. Soon afterward, however, other researchers were producing entire life forms (not just cells or tissues) that did similar things, including bacteria, fish, and even mice that glowed in the dark. And then, in 2000, Chicago artist Eduardo Kac, in collaboration with a French molecular geneticist living in Paris, took an albino rabbit named Alba and gave its white fur strands a fluorescent green hue by inserting into the rabbit's genome the gene for green fluorescent protein (GFP), which causes certain species of jellyfish to light up in the presence of ultraviolet light.

The rabbit Alba was now a transgenic animal, created, according to Kac, not as a research tool but as an "art object." It is stunts like this that give something of a bad name to recombinant DNA technology. Alba was viewed as an "unnatural" animal, and this was offensive to many people, including animal rights activists and even some scientists.

But, in the same year, a molecular genetics team at Cambridge University in England imported the same jellyfish GFP gene into potatoes, producing a plant whose leaves glowed "like a white shirt in a disco," according to the team leader, Anthony Trewavas. This was not just another molecular acrobatics trick or a form of weird living sculpture. The potato leaves shone only when the plant needed water. Water being an increasingly scarce and expensive commodity in an age of climate change and prolonged droughts, a potato that tells you when it needs watering could be a useful commodity. This is an example of a recently emerging and potentially valuable branch of genetic engineering technology, the creation of "climate-change-ready" crops.

But, as apparently unnatural as the glowing tobacco plants, Alba the fluorescent rabbit, and the shining potato leaves might have been, were they really any more extraordinary than the grafted-into-existence pomato (the Tomtato)? Grafted interspecies organisms are *themselves* chimeras, after all, mixtures of genes from two different species. One of the first recorded graft chimeras goes well back into history. Produced in Florence in the early seventeenth century, the Bizzaria contains characteristics of both of its parent organisms, the Florentine citron and the sour orange, expressed together in the same fruit. In some cases the characteristics of the two fruits are blended, but in other cases not.

The plant was called the "Bizzaria" because of the freakish structure and appearance of some of the fruits, which were half citron, which is green, and half sour orange, which is yellow. In other cases the yellow-green pattern alternated in sections, like the stripes on a hot air balloon. Horticulturists had a hard time even describing this particular chimeric fruit, using adjectives like "extremely curious," "fantastic," and "anomalous." A 1902 textbook described the Bizzaria as "a compound fruit, with the two kinds either blended together, both externally and internally, or segregated in various ways." In one case, "five longitudinal stripes of the color of a Citron were interpolated in the fruit of the Orange. Other fruits were, on the whole, like Oranges, excepted as regarded an eighth of their mass, which in form, color, and taste resembled a Citron, and was also peculiar for its extreme convexity."

In other words, it didn't take recombinant DNA methodology to produce chimeric Frankenfruits. Nature alone, acting all by itself, is quite capable of accomplishing the feat. Indeed, the most surprising aspect of the Bizzaria is that this graft-chimera had not been produced intentionally, by human agency, but was rather the product of an adventitious graft of a sour orange shoot onto a citron rootstock. The strange combination fruit that resulted was discovered in 1640 by the Florentine gar-

The Bizzaria. Labrina, "Citrus aurantium 'bizzarria' frutto acerbo," *Wikimedia Commons*, November 2, 2011, by Labrina—own work, CC BY-SA 3.0, https://commons.wikimedia.org/w/index.php?curid=19015504.

dener Pietro Nati, who in his book, *De malo limonia citrata aurantia vulgo la bizzarria,** said that the weird citron-orange fusion had been created "by no modification of the fruits, by no ingenuity in handling; by no skill in crossing; but solely as an act of nature."

The moral of the story is that recombinant DNA techniques are not unique in their ability to produce bizarre, anomalous, or even freakish plants, fruits, or animals. Nature by itself is quite up to the task. And so are the other, conventional methods of modifying genes.

■ All the methodologies used to create new plant varieties have their own limitations, uncertainties, and mysteries. This is true even of the simplest and most conventional breeding technology, crossing. Richard Lewontin, the Harvard geneticist, has explained that "introducing genetic variation by crossing between organisms is imprecise. A cross between two varieties is indiscriminate in the hereditary characteristics that are transmitted. Thus if one attempts to introduce disease resistance into an especially high-yielding variety of wheat by crossing that variety with one that has the disease resistance but not the high yield, the result will be a variety with improved resistance but lower yield."

* Of the Evil Citron-Orange Lemon, Called the Bizzarria.

When two plant varieties or species are crossed, a mixture of entire genomes takes place, in a process that is uncontrolled and whose precise outcome is often unknown in advance. Radiation breeding, chemical mutagenesis, and space breeding methods, much like the natural production of sports or spontaneous mutations, are also essentially random processes, with outcomes generally unpredictable beforehand.

But, surprisingly in view of its antiquity as a practice, it is grafting—the very method that produced the Bizzaria—that is the most mysterious breeding method of them all, far more so than recombinant DNA technology, which is well understood from the nucleotides up. How grafting works on the structural or morphological level is reasonably clear. At the point where the graft occurs—the graft junction—cells rupture, collapse, and adhere to the opposing tissue. Over time, the cells interdigitate, interlocking like the fingers of two hands folded together. Cell division at the junction point produces a mass of pluripotent cells (those capable of giving rise to cells of different types) to form callus tissue, in a sort of wound-healing process that occurs in plants. It is also known that all three of a plant cell's genomes can be transferred by grafting: those in the nucleus, in the chloroplast (the organelle in which photosynthesis takes place), and in the mitochondria (energy-producing organelles).

But what then happens further down, at the molecular level, to yield the new, hybrid plant is not well understood even today. It's so unexplained, in fact, that scientists attempting to communicate this lack of knowledge about the subject find themselves using language not normally seen in specialist academic journals. For example, a comprehensive review of the mechanisms of grafting, published in 2014, concluded that "this plant propagation practice is still shrouded in mystery." And in a later (2015) study of plant grafting, the authors bluntly asserted, "To date the molecular mechanisms of graft formation remain unknown," and, furthermore, "how grafting evolved remains an unresolved question."

But, despite these considerable unknowns, the fruits and vegetables that are products of grafting are viewed as risk-free and entirely unproblematic by the general public—as well as by opponents of GMOs. And rightly so, since no adverse health consequences have been attributed to them or to the process by which they were produced. But, then again, no adverse health consequences have been attributed to GMOs, either. Which means that, on the specific criterion of causing health problems, these two methods of genetically modifying organisms are on a level with each other.

But if that is true, then the question arises as to why GMOs produced by recombinant technology are singled out for special treatment, precautionary handling, and strict government regulation. Recombinant molecular techniques constitute just one method among many that yield new varieties of plants and animals. Yet GMO critics claim that varieties produced by recombinant methodologies are special. They are rife with unknowns (as are all the other methods of genetic modification), risky, and potentially unsafe—particularly unsafe to eat.

But is there any rational basis for such claims?

SAFE TO EAT?

central paradox of the GMO safety issue is the radical mismatch between the zealousness of the opposition to GMOs and the empirical record of their harmlessness over the entire lifespan of their successful use as foods. The nuclear power industry has had its Three Mile Island meltdown, its Chernobyl disaster, and its Fukushima accident. The chemical industry has had its Bhopal catastrophe. And the GMO industry has had its . . . nothing at all. There have been no major or minor GMO disasters, mini-catastrophes, or indeed any documented instances of human health being compromised by the consumption of genetically modified foods. GMOs have not laid waste to the environment or damaged it in any significant way. So far as is known, not a single person, alive or dead, has been harmed by any genetically modified food over the entire quarter century since the first GMO to be marketed, the Flavr Savr tomato, was offered for sale, bought, and eaten by consumers in 1994.

Since that time, millions of acres of farmland have been planted in genetically modified crops, reaching a peak of 448.5 million globally in 2014, principally in corn, soybeans, cotton, and canola. Corn is the most widely grown crop in the United States, and fully 88 percent of current corn production (and 95 percent of the nation's sugar beets) are of genetically modified varieties. Most of the GMO corn is used as animal feed, but some is present in processed foods such as cereals. And, in Canada, much of the sweet corn consumed by humans is genetically modified. Americans and Canadians have been eating genetically engineered foods since 1995, with no known ill effects.

Some critics of genetically modified foods have claimed that this is because the proper studies have not been done. In 2013, for example, the

European Network of Scientists for Social and Environmental Responsibility issued a position paper which said, "There are no epidemiological studies investigating potential effects of GM food consumption on human health." That was true, but there are two good reasons for the absence of such studies. The first is that epidemiology is the study of the origin, incidence, transmission, and spread of human diseases. Since no diseases have been attributed to GMOs, no epidemiological studies of them are possible. The second reason is that, since by definition "*potential* effects" do not yet exist, it is not possible to study them. There are simply no illnesses, diseases, or effects to investigate.

That said, in 2016 the US National Academy of Sciences published a study of genetically modified foods in which it considered epidemiological datasets from the United States and Canada, where genetically engineered foods have been consumed since the mid-1990s, and similar datasets from the UK and Western Europe, where such foods have not been widely consumed. The Academy of Sciences concluded, "No pattern of differences was found among countries in specific health problems after the introduction of GE foods in the 1990s."

Yet, even in the face of their remarkably spotless track record, the opposition to GMOs on the part of their fiercest critics can be accurately described as extreme, even feverish. This negative critical reaction is practically unequaled in the history of technology, comparable only to the Luddite hostility to and seizure and destruction of textile machinery during the early 1800s. What accounts for this bizarre polarity between the militant emotional opposition to GMOs on the one hand and their unblemished, 25-year-long safety record on the other? Why do critics regard GMOs as the Antichrist?

■ Critics despise GMOs largely because their views are based more on *perceptions* than reality and because there are multiple sources of negative perceptions about genetically modified foods. One source is a legacy of the Precautionary Principle and the matrix of regulations, restrictions, prohibitions, and other obstacles that followed from its incorporation into legal frameworks governing GMO research, field trials, distribution, and use as foods. From the fact that a multiplicity of stringent regulations is in place it is an easy, almost inevitable, inference that organisms so regulated must be extremely risky, bristling with unknown dangers, and likely to cause harm to human beings and the environment. After all, where there's smoke . . .

Another source of negative perception is the fact that starting in 2001 the European Union imposed a de facto moratorium on new GMO approvals. In addition, by 2015, 17 of the European Union's 28 member states had imposed bans on the cultivation of GMO foods. While that was a clear majority, it was nevertheless not a continent-wide ban. Spain, for example, resolutely continued to grow its own genetically modified maize. But the moratorium on new approvals coupled with the 17-nation prohibition, gave rise in the popular consciousness to the blanket, albeit false, view that GMOs were "banned in Europe" as a whole. Obviously, no sensible state authority, or combination of them, bans products that are safe. GMOs, therefore, must be unsafe. That, at least, was the *perception*.

Standing against that perception was the fact that although most European nations banned the *cultivation* of GMO crops, every European nation nevertheless *imports* them for use as livestock feed. Indeed, as a whole, Europe imports more than 30 million tons of GMO corn and soy each year, making the European Union one of the world's largest regional consumers of genetically engineered crops.

It is important to recognize in this connection that the European moratorium and prohibitions are political rather than scientific measures. That is, they represent laws passed by the European Parliament, most of whose members are not scientists. In fact there is a striking contrast between the laws passed by parliaments and extensive studies of genetically modified organisms by the scientific bodies within those same European Union states that have concluded that GMO foods are at least as safe as non-GMO foods (table 1).

A later (2015) study showed that, on a worldwide basis, some 280 organizations and scientific institutions have concluded on the basis of evidence that GMOs are as safe to eat as conventional foods. "Interestingly a large part of these institutions are located in Europe, the conti-

Table 1. List of European and international scientific institutions
that have concluded genetically modified crops are safe to human health
and the environment and that the technology poses no inherent risk

Institution	Country	Year
Nuffield Council on Bioethics	UK	1999
European Research Directorate	European Commission	2001
French Academy of Science	France	2002
French Academy of Medicine	France	2002
Director-General, WHO	International	2002
International Council for Science	International	2003
Royal Society	UK	2003
United Nations Food and Agriculture Organization	International	2004
British Medical Association	UK	2004
German Academies of Science and Humanities	Germany	2004

Source: Adapted from Adrian Dubock, "The Politics of Golden Rice," *GM Crops & Food* 5 (2014): 210–22.

nent that has put more obstacles to the commercialization of these crops," the author said.

But if all of these European scientific organizations (as well as their counterparts in the United States and elsewhere) have concluded that GMOs are safe to eat, then why does the perception persist in people's minds that, since the foods are allegedly "banned in Europe," they must be unsafe? Lisa Weasel, a molecular biologist at Portland State University and author of the book *Food Fray: Inside the Controversy over Genetically Modified Food* (2009), attributed the persistence of these negative perceptions to the "cognitive miser" principle of decision making. Essentially, this is a mode of reasoning "in which conclusions are drawn from only the most basic and minimal information, collected from the most convenient and readily available sources."

For cognitive misers, the belief that GMOs are "banned" across an entire continent overrides and carries more epistemological weight than the conclusions of scientific organizations about something as arcane, complex, and remote from ordinary experience as the genetic modification of organisms.

A further source of negative perceptions regarding GMOs is opposition to genetically modified foods on the part of activist organizations such as Greenpeace and Friends of the Earth that mount street demonstrations against GMOs, issue position papers opposing genetically engineered crops, and engage in campaigns against allegedly rapacious multinational biotechnology companies that the activists portray as somehow forcing GMO foods down the throats of consumers, risking their health and hurting the environment in the process. The organic food industry also disparages GMOs while presenting the consumption of organic foods as healthy, environmentally friendly, and even morally virtuous.

In consequence of these portrayals, being opposed to GMOs becomes an act of "virtue signaling," which also contributes to a negative perception of genetically modified foods. Virtue signaling is defined as "the action or practice of publicly expressing opinions or sentiments intended to demonstrate one's good character or the moral correctness of one's position on a particular issue." Hostility to GMOs—as announced on "GMO/OMG" T-shirts, buttons, bumper stickers, placards, posters, and banners—is then largely an exercise of public relations or advertising, an exhibition of one's socially responsible and ideologically correct views on the subject, irrespective of the facts of the case and the weight of scientific evidence to the contrary.

Negative perceptions of GMOs are reinforced by several other influences. Consider, for example, the mass destruction of GMO laboratory experiments and equipment, greenhouses, and outdoor field trials, which was almost a rite of passage among European activists starting in the 1990s. In 2012, Marcel Kuntz of the Université Joseph Fourier in Grenoble, France, compiled a listing of 80 documented acts of vandalism that had been committed in Europe between 1997 and 2011 against government and academic GMO research projects, often by unknown perpetrators.

His tally, Kuntz reported, "is a vast under representation of the total number of GMO field destructions that have taken place across the EU, since destructions of trials implemented by private companies are not listed here. Germany alone has had more than 100 acts of vandalism."

This series of crimes, which were widely reported in the media (but in many cases were not vigorously prosecuted by the local police or the courts), created a climate of fear and distrust pertaining to any aspect of GMO research, experimentation, and safety testing, further poisoning the already highly tainted emotional atmosphere surrounding GMOs.

People's perceptions are further shaped by value-laden words and images, of which GMOs are saddled with more than their fair share, beginning, of course, with the potent, unforgettable, and by now inescapable term of abuse, "Frankenfood." This handy, go-to catchword was coined in 1992, in a letter to the editor of the *New York Times*, which read in its entirety:

> "Tomatoes May Be Dangerous to Your Health" (Op-Ed, June 1) by Sheldon Krimsky is right to question the decision of the Food and Drug Administration to exempt genetically engineered crops from case-by-case review. Ever since Mary Shelley's baron rolled his improved human out of the lab, scientists have been bringing just such good things to life. If they want to sell us Frankenfood, perhaps it's time to gather the villagers, light some torches and head to the castle. PAUL LEWIS Newton Center, Mass., June 2, 1992

But, by that time, many European villagers had long since lit their torches and stormed, if not their castles, then at least their nearest GMO lab, greenhouse, or field trial.

The equally indelible, canonical, and thought-stopping graphic image was that of scientist Julianne Lindeman, of Advanced Genetic Systems, spraying genetically altered "ice-minus" bacteria over a patch of strawberry plants in a field in near Brentwood, California, in 1987. These GMO bacteria had been designed to protect the fruit from frost, and the organism in question (a strain of *Pseudomonas syringae*) was the very first GMO authorized for open release into the environment. For this historic experiment, the California Department of Food and Agriculture required that the scientist doing the spraying be clad in an outfit more appropriate for biosafety level 4 work on virulent pathogens such as Ebola viruses or anthrax spores. Accordingly, Lindeman was garbed in a white plastic "moon suit," complete with hood, gloves, goggles, and a respirator that looked like a gas mask capable of protecting the wearer from a chemical or biological warfare attack. All the while a group of photographers and reporters stood by 10 feet away clad in ordinary street clothes. A photograph of the moon-suited scientist, which was widely published in newspapers at the time, became an iconic symbol among anti-GMO activists.

Banned in Europe. GMO/OMG. Field trials vandalized. Frankenfoods. Moon suits, respirators, gas masks. What chance did an innocuous GMO have in the face of such unremittingly hostile and negative PR?

■ But that series of GMO-tarnishing hexes, symbols, and other evil influences by no means exhausted the inventory of negative-perception generators. Further sources derived from a number of urban legends, memes, and myths about GMOs and their effect upon the environment. There are far too many of these folk tales, misconceptions, and distortions to address individually and systematically in short compass, but one in particular may be taken as a representative case to illustrate the ways in which these fictions originate, spread, and persist despite abundant evidence of their falsity.

One of the greatest of all of these urban legends is the common and persistent belief that GMOs are killing off the monarch butterfly, driving it to the brink of extinction. It must be noted first of all that the monarch butterfly is an example of a rare if not unique zoological phenomenon: a charismatic insect, a bug with all the cuddly aesthetic and emotional appeal of a panda bear, baby seal, or puffin. Unequivocally, it is the Bambi of the insect world. And in 1999 a team of Cornell University entomologists performed an experiment in which they raised monarch larvae (which is to say, butterflies in the caterpillar stage of their life cycle) on milkweed leaves that the experimenters had dusted with pollen from *Bt* corn.

Bt corn is a genetically modified corn that contains genes from the common soil bacterium *Bacillus thuringiensis* (*Bt*). The bacterium accumulates crystals that are naturally toxic to moths and butterflies, a fact that was first discovered in 1901 by a Japanese biologist who found that the bacterium was responsible for killing large populations of silkworm moths in the larval phase. Ten years later, in 1911, the same bacterium was independently "rediscovered" by a German scientist living in the town Thuringia, where the bacterial toxin was killing Mediterranean flour moths, hence the name.

Bacillus thuringiensis is therefore a natural pesticide—a biopesticide—and farmers started using *Bt* spores, and crystalline proteins produced by the bacterium, as insecticides as early as 1920. Because the *Bt* toxin is a natural substance and not a synthetic chemical, it is especially favored by home gardeners and organic farmers, who apply it to their plants in the form of powders and sprays. And it is so environmentally friendly (it breaks down in water and under the ultraviolet rays of the sun) that, in her book *Silent Spring*, Rachel Carson, the pioneering environmentalist, went so far as to praise its use as a sustainable alternative to chemical pesticides.

"Shortly after eating foliage covered with this toxin, the larva suffers paralysis and dies," Rachel Carson wrote. "For practical purposes, the fact that feeding is interrupted promptly is of course an enormous advantage, for crop damage stops almost as soon as the pathogen is applied." For Rachel Carson, *Bt* toxin was something benign, environmentally friendly, and good.

Recombinant DNA technology later made it possible to introduce the gene that coded for the *Bt* toxin into the very plants themselves. This would in effect produce a plant that has inside itself its own inherent natural defense against certain predators. Farmers wouldn't have to dust or spray the toxin; the plant would now already have it as a built-in feature. It is worth noting that many plant species had acquired such built-in defenses against predators naturally, through evolution by natural selection.

In 1995, *Bt* corn was registered by the EPA for use in the United States, and in 1996 the EPA approved it for full commercial use against the European corn borer, a pest that in some years was costing American farmers an estimated $1 billion in crop damage, equivalent to a loss of about 20 bushels of corn per acre.

Given the rapid action of this natural toxin, it was no surprise that, when the Cornell entomologists raised monarch larvae on milkweed leaves liberally covered with *Bt* corn pollen, many of the larvae died, while the survivors suffered various other ill effects. After all, *Bt* corn expressed a protein that was specifically chosen to kill the caterpillars that otherwise would be feasting on the corn, and the pollen evidently contained at least trace amounts of the same toxic substance.

The Cornell researchers drew a rather alarming conclusion from their findings. In a one-page letter to *Nature*, "Transgenic Pollen Harms Monarch Larvae," published in May 1999, they wrote, "These results have potentially profound implications for the conservation of monarch butterflies. Monarch larvae feed exclusively on milkweed leaves." The reader is left with the impression of imminent mass death among monarch larvae feeding on milkweed, their only food, sprinkled with lethal quantities of *Bt* corn pollen.

This was instantly front-page news in the *New York Times* ("Altered Corn May Imperil Butterfly, Researchers Say," May 20) and a major news story around the world. GMO opponents were quick to invent and adopt the rhetoric of "killer corn" and to make the most of this looming, dire potential threat to the monarchs.

Thus was born the "GMOs are killing off the monarchs" meme.

Other scientists, however, soon pointed out serious flaws with both the Cornell experiment and with the researchers' interpretation of their own results. First was the fact that the lab study did not accurately reflect real-world conditions in a field planted in *Bt* corn. Specifically, there were higher concentrations of *Bt* corn pollen on the laboratory milkweed than found in typical agricultural settings. Second, the lab larvae were restricted solely to leaves that had been covered with *Bt* corn pollen (the larvae had been placed on the leaves by hand, five to a leaf), whereas in the open field the caterpillars can avoid pollen-covered corn and feast on milkweed that is clean and wholesome. Third, the monarch caterpillars would have suffered similar (if not worse) fates had the corn not been dusted with *Bt* corn pollen but rather with the *Bt* toxin itself, the very substance whose use against insect pests had been endorsed by none other than Rachel Carson. This meant that GMOs per se were not the problem; it was the *Bt* toxin itself that was the problem, irrespective of how it was applied.

The scientists who made these and numerous other corrections produced a volume of data and analysis that was sufficient to fill six peer-reviewed papers on the relatively narrow subject of *Bt* corn pollen deposition on milkweed and its effects on the monarch larvae. The six papers were published together in an issue of *PNAS*, the *Proceedings of the National Academy of Sciences*. The last of them, a wrap-up study that summarized the results of an enormous collaborative research effort by scientists in several states and in Canada, reached the final conclusion that, "in most commercial hybrids, *Bt* expression in pollen is low, and laboratory and field studies show no acute toxic effects at any pollen density that would be encountered in the field. Other factors mitigating exposure of the larvae include the variable and limited overlap between pollen shed and larval activity periods, the fact that only a portion of the monarch population utilizes milkweed stands in and near cornfields, and the current adoption rate of *Bt* corn at 19% of North American corn-growing areas. This 2-year study suggests that the impact of *Bt* corn pollen from current commercial hybrids on monarch butterfly populations is negligible."

The body of evidence marshaled by these researchers should have laid to rest forever the meme that GMOs were depopulating the monarchs. But an accident of timing blunted its effectiveness. The papers were published in the October 9, 2001, issue of *PNAS*, barely a month following the September 11 terrorist attacks and the anthrax-letter attacks a week later, both of which filled media accounts to the exclusion of good news about butterflies.

But there was yet a second chapter to the story. Later, in 2012, two Iowa State University scientists advanced a new, different, and original theory about what was killing off the monarchs. The culprit was not GMOs but rather chemical herbicides. In their study, "Milkweed Loss in Agricultural Fields because of Herbicide Use: Effect on the Monarch

Butterfly Population," the authors attributed adverse effects on the monarchs to the use of the herbicide glyphosate, better known under the trade name Roundup.

Glyphosate is neither a GMO nor is it produced by GMOs. It is a synthetic chemical, produced in a series of complex chemical reactions in large industrial vats. In other words, according to this new account it is *synthetic chemicals*, not GMOs, that are wiping out monarchs. But this exact point, which is neither very subtle nor especially hard to understand, was lost on or ignored by a number of environmentalist publications whose news accounts of the glyphosate hypothesis nevertheless continued to propagate the original and erroneous GMO meme. And so when the magazine *Mother Jones* reported on the scientists' new synthetic chemical theory, its article was headlined, "Researchers: GM Crops Are Killing Monarch Butterflies After All." A similar story in *Grist* was headlined: "Study: GMO Crops Are Killing Butterflies."

Both claims were flatly and demonstrably untrue, but they nevertheless fit the anti-GMO narrative line better than the actual facts, which didn't. And so the GMO-monarch meme lives on to this day.

■ Still, beyond the myths, memes, distortions of fact, and outright falsehoods—along with the reverse halo effect that has long bedeviled the subject of genetically modified organisms—the actual record of their safe use as foods is a blunt empirical fact that calls out for explanation. What explains it?

First there is the fact that the amount of genetic alteration made in the genome of any given organism is tiny compared to the size and complexity of its genome as a whole. For example, the key genetic change made in producing the first GMO to be marketed, the Flavr Savr tomato, was the insertion of an "antisense" gene into the tomato genome. An antisense gene is essentially a scrambled gene, one that makes no sense to the cell's protein synthesis machinery; it is analogous to writing the letters of a word upside down and backward so that they are unintelligible to a human reader. In the case of the Flavr Savr tomato, creating the antisense gene was a simple matter of rearranging a handful of nucleotides.

As one of the scientists involved in the development of the Flavr Savr, Belinda Martineau, herself put it, "The genetic modification itself—the addition of a relatively small piece of DNA containing a known, albeit 'antisensed,' gene from the same organism—was minimal."

Indeed it was. The genome of the tomato was sequenced in 2012 by a consortium of scientists who discovered that it contains 31,760 genes. That is a lot of genes, especially in view of the fact that it is larger than the entire genome of a human being. The introduction of a single, short antisense gene into the Flavr Savr tomato had the limited effect of reducing the expression level of the enzyme that is responsible for ripening. In other words, the modified plant is identical to its parent plant with the exception of the single desired trait that has been incorporated.

This type of minimal alteration of the genome is typical of the genetic modification of organisms. In most cases, only a small number of genes are introduced, and both the genes and their functions are well known and well characterized. By contrast, many types of conventional breeding—including radiation breeding, chemical mutagenesis, and grafting—introduce changes to the genome that are unknown and that may in fact represent far more extensive modifications to the original genome—on the order of hundreds or thousands of uncharacterized genes—than the small number of known genes added by molecular methods.

Alan McHughen, a molecular geneticist and GMO developer currently at University of California, Riverside, has noted:

> In mutation breeding we have no idea what genetic changes have been induced and little or no molecular characterization of even the evident novel trait. In rDNA [recombinant DNA] breeding we know exactly what genes have been introduced and have them fully characterized, right down to the molecules.
>
> If mutation breeding is acceptable and safe with the current level of regulatory scrutiny, then there is no logical basis to categorically "draw the line" between rDNA and mutation breeding.

However true that may be, GMO critics have nevertheless claimed that, for all its precision in the molecular characterization of DNA inserts, there are significant ways in which the recombinant DNA process is *less* precise than is commonly advertised by its proponents. One in particular is that no matter what the method of insertion—that is, whether by particle bombardment or by *Agrobacterium*-mediated transformation—genetic engineers have no control over, and normally do not know, where in the recipient genome the introduced sequences are located. They are inserted essentially at random.

In addition, the insertion of a transgene sometimes causes unintended effects in the chromosomes where the insertion occurs. In their review article, "The Mutational Consequences of Plant Transformation," Jonathan R. Latham and two other authors stated, "Transgene insertion is infrequently, if ever, a precise event. Mutations found at transgene insertion sites include deletions and rearrangements of host chromosomal DNA and introduction of superfluous DNA."

In consequence, "mutations may result in transgenic plants with unexpected traits. Despite the supposed precision of genetic engineering, it is common knowledge that large numbers of individual transgenic plants must be produced in order to obtain one or a few plants that express the desired trait in an otherwise normal plant."

According to these and other GMO critics, then, such problems raise significant questions about the safety of genetically modified foods.

■ The problem of unintended effects in genetically engineered crops has been one of the most frequent criticisms leveled against GMO foods. And, as one might expect, that problem is also one of the most frequently addressed by scientists and GMO advocates. Under these circumstances, one example will have to stand as representative of what amounts to a consensus of scientific thought on the subject.

A comprehensive review of the unintended effects problem was produced by a group of 14 European scientists who in 2004 published in the journal *Food and Chemical Toxicology* a study, "Unintended Effects and Their Detection in Genetically Modified Crops." In this analysis, the authors made the following points. First, in regard to the chromosomal rearrangements caused by transgene insertion, integration of an insert

into a plant genome is subject to the same natural DNA repair mechanisms that are used by cells to repair normal, untransformed DNA.

"The major source of natural variation and of breeding programmes
is the natural molecular mechanisms of DNA exchange and repair," the
authors wrote. "These mechanisms are the same for all crops, irrespective of whether the DNA has been specifically modified by genetic engineering techniques or has been altered via conventional crossing of different varieties." Genomic alterations that occur naturally are repaired
naturally, and the same is true of alterations that are introduced by molecular methods.

Second is the important fact that unintended effects occur in *all*
forms of plant breeding, whether made by traditional or by molecular
methods. In other words, the production of unintended effects is not a
new complication introduced by recombinant methodology but is one
that breeders have long been aware of and for which they have developed and adopted corrective measures. "In classical breeding programmes, extensive backcrossing procedures are applied in order to remove unintended effects."

For example, a given breeding program might start with the crossing
of 1,000 individual plants that constitute the parental generation (P1).
That initial crossing of those parent plants yields the first filial generation
(F1). The F1 generation is evaluated "by eye," a "low-grade" type of selection, to identify promising variants, if indeed any have been produced.

"It is possible that with some parents the resulting F1 generation produces no lines with obvious desirable traits," said the authors. "The selection of starting parental material is therefore paramount and is an
iterative process requiring years of experience dealing with diversity in
available genetic resources."

Any plant lines that do exhibit desirable traits are then crossed again,
to produce a second filial generation (F2), which breeders evaluate for
properties such as general morphology and disease resistance. Classical
breeding involves several additional filial generations, with further,
higher types of selection criteria used to screen for such traits as yield,
fruit or seed quality, and so on.

It often happens in traditional breeding that the trait bred for produces instead an unintended effect. For example, a line of barley bred for mildew resistance turned out to have low yield as an unintended effect. And a variety of potato bred for pest resistance in the end had an unacceptably high glycoalkaloid content, another unintended consequence.

The same strategy of repeated backcrossing that is used in traditional plant breeding to remove unintended effects is also used on plant lines developed by molecular methods. In addition, "during the development of transgenic plant varieties and for any given trait(s), a large number of transformants/clones that do not perform up to required expectations will be discarded through assessment in the laboratory, glasshouse, and small scale field trials. In all cases, new cultivars produced by genetic engineering are extensively tested and screened prior to commercial release."

It is important to note that while GMO critics consistently assume that unintended effects are harmful, or at least potentially dangerous, the fact is, the authors write, "unintended effects do not automatically imply a health hazard. Unintended effects may have absolutely no impact on health, or may even be beneficial by reducing potentially toxic substances."

The overall conclusion reached by this group of scientists was that "there is no indication that unintended effects are more likely to occur in GM crops than in conventionally bred crops. Unintended effects may have positive, negative, or indeed no consequences on the agronomical vigour or safety profile of the crop. The same field selection processes apply to both conventional and GM breeding. This selection process takes many years and removes major unintended effects."

■ As there is a paradox concerning the hostility of critics toward GMO foods despite their record of safe use, so there is a corresponding counter-paradox concerning commercially available, unregulated, *non-GMO* foods that are nevertheless known to be *unsafe*, at least in their effects on some consumers. Many conventional foods contain known toxins and allergens, the most familiar example being peanuts, to which some children (estimated at 0.6 percent of the American population) are so allergic that eating them will cause a range of ill effects including, at the extreme, even death, unless the hormone epinephrine is promptly administered by injection.

In 2016, author Ali Jaffe wrote a *New York Times* story, "How 12 Epi-Pens Saved My Life." EpiPens are devices containing injectable epineph-rine to alleviate this kind of anaphylactic shock reaction.

> I used my first EpiPen at the age of 7, after eating chocolate cake at a restaurant, sliced with the same knife used on a walnut cake.
>
> Five years later at a sleep-away camp in New Hampshire, it was the marshmallow fluff, contaminated by peanut butter. The nurse was so nervous she stabbed herself (EpiPen No. 2) instead of me.
>
> I'm 24 now. All told I have had to use an EpiPen 12 times in the past 17 years for life-threatening allergic reactions to nuts.

Other people are allergic to foods as various as milk, eggs, soy, wheat, and shellfish. All of these foods are freely available everywhere and with-out restriction or labels warning of possible adverse health consequences, including death. But if any genetically modified food had produced even the very slightest allergic reaction in a human subject at any level of expo-sure, that food would be banned from the marketplace without delay.

These examples suggest that safety is a relative, subjective property since no food is safe for all people in all circumstances. The examples fur-ther suggest that double standards are again at work in assessing the safety of conventional versus GMO foods. Conventional foods are regarded as "safe" and are freely available and unregulated while nevertheless posing known health hazards. GMOs, by contrast, are held by critics to be unsafe not because of known, proven, suspected, or probable hazards but rather because they are claimed to present merely potential, hypothetical, and even "unknown risks" to human health or the environment.

Double standards also apply to the amount, nature, and stringency of the regulatory requirements that must be satisfied by conventional versus GMO foods. The number of government agencies overseeing the food supply is considerable, as is the weight of regulatory apparatus even in such GMO-friendly countries as the United States or Canada. Not only do we have the FDA, USDA, and EPA, agencies familiar to most people, but also agencies, commissions, regulatory bodies, and codes that are largely unheard of by the average consumer. The little-known and obscure Codex Alimentarius is one example of the latter.

Codex Alimentarius, Latin for "food code," is a collection of international standards, guidelines, and best practices that have been advanced and adopted by the Codex Alimentarius Commission, a body established jointly by the United Nations Food and Agriculture Organization and the World Health Organization in 1961. This was well before the advent of biotechnology, recombinant DNA, and GMO foods, and so the codex guidelines were originally developed to protect the safety and integrity of the world's supply of conventional foods. Later those standards were extended to cover GMO foods as well.

Most of the world's countries, including the United States, Canada, the United Kingdom, as well as the nations that make up the European Union, are members of the codex and are committed to upholding its standards. But the list of standards, regulations, and requirements that conventional foods must meet are dwarfed by those imposed on genetically modified foods, not only by the Codex Alimentarius but also by each nation's regulatory agencies. To get a realistic picture of what this means in real-world practice, it may be useful to consider a case of the radically different regulatory requirements that have to be met by a new plant type that is a product of conventional breeding methods as opposed to those imposed upon a GMO variety of the very same plant.

In the 1980s Alan McHughen, then of the Crop Development Center at the University of Saskatchewan, developed and field-trialed a new high-yield, early-maturing variety of linseed. Linseed is the seed of the flax plant and is an important crop in Canada. It is the source of linseed oil, used primarily in paints and varnishes, but it is also marketed as an edible oil and as a nutritional supplement.

McHughen produced his new flax variety by means of a recently invented breeding technique known as somaclonal variation breeding, a process by which whole plants are regenerated not from seeds but from the tissue cells of a parent plant. Although the technique is relatively new,

Pile on left: documents required in support of the application for the GMO flax variety CDC Triffid, which was approved for Canadian release only. *File folder on right*: documents supporting the non-GMO flax variety CDC Normandy, approved for worldwide distribution, cultivation, and sale. © Alan McHughen. Used with permission.

the somaclonal variation method is nevertheless considered a form of "conventional" breeding since it does not involve the direct molecular manipulation characteristic of genetically modified organisms.

By 1995 McHughen's new linseed had been accepted by Canada's Variety Registration Office and was approved for worldwide distribution, cultivation, and marketing under the trade name CDC (Crop Development Center) Normandy. The documents required to obtain these approvals fit neatly inside a single thin file folder.

Later, McHughen also developed a second, genetically modified flax plant, one that was tolerant of certain herbicides. This new, GMO version was a variety called CDC Triffid.

When CDC Triffid was finally approved—but for release solely in Canada, not worldwide—the stack of required documents was more than 2 feet tall. This impressive tower of paper represents many years of work, lots of money, and all but endless types of molecular characterizations, experimental data, trials, and tests. It may be taken as a visual index of the huge obstacles and delays that faced not only CDC Triffid but GMO crops in general, and in particular Golden Rice.

■ An additional and important part of the explanation of the safety record of GMO foods derives from the fact that DNA is largely, and in most cases totally, destroyed in the process of digestion. This is true whether the DNA is "natural" or recombinant.

Several experiments have established this point. For example, in their 2004 study, "Assessing the Survival of Transgenic Plant DNA in the Human Gastrointestinal Tract," a group of British researchers used a novel experimental strategy to discover the fate of DNA at two separate stages in the digestive process. They started with a group of 19 volunteer subjects, 7 of whom had undergone resection surgery on their small intestines and had part of their upper intestinal tracts removed, with the result that their digestive end products were diverted from the body into a colostomy bag. The other 12 subjects had fully intact digestive systems.

All of the volunteers were fed a meal containing genetically modified soy that contained a known transgene, and after several hours the digestive products of all the test subjects were assayed for the presence of transgenic DNA. It turned out that in the 7 colostomy subjects, a small portion of the transgene (a maximum of 3.7 percent of the full gene) survived passage. In the 12 subjects with intact digestive tracts, no DNA was recovered. In other words, partially digested food retained a small portion of the original transgene, but fully digested end products showed no trace of the DNA at all.

On this basis, the researchers concluded, "We have shown that a small proportion of the transgenes in GM soya, like native soya DNA, survives passage through the human upper gastrointestinal tract but is completely degraded in the large intestine. . . . Thus, it is highly unlikely that the gene transfer events seen in this study would alter gastrointestinal function or pose a risk to human health."

These results, which have been independently replicated by other studies on human and animal subjects, as well as by in vitro laboratory experiments with simulated gastric juice, are not surprising. Despite the adage that "You are what you eat," humans do not become the plants and animals that we consume as food. As one researcher put it, "If gene transfer from plants to humans or animals were commonplace, we should all be green by now." That we aren't is explained by the fact that DNA of any type is broken down when it is used as food, beginning in the processes of preparation and cooking and continuing during the course of digestion.

As is common knowledge, DNA is composed of the four nucleotides, A, T, G, and C (adenine, thymine, guanine, and cytosine), which are held together along each strand by strong covalent bonds, while the two strands are held to each other by relatively weak hydrogen bonds. In foods that are cooked, DNA breakdown begins when the weak hydrogen bonds are broken as water is heated near to the boiling point (ca. 201–208 degrees Fahrenheit). This heat denatures the DNA, causing the two strands to unwind and separate.

Additional DNA degradation occurs chemically, beginning in the mouth. Saliva contains the enzyme deoxyribonuclease, which cuts some of the bonds between nucleotides, separating the strands into smaller pieces. Recent (2015) experiments show that degradation continues in the stomach, caused not so much by the action of stomach acids but rather by means of another component of human gastric juice, the enzyme pepsin, which breaks the nucleic acid molecule into yet smaller fragments. Other enzymes in lower parts of the digestive tract essentially finish the job of DNA destruction, inactivating the molecule.

The process is not perfect, however, and it is known that some DNA survives or escapes all these mechanisms and passes through the intestinal wall to enter the bloodstream. GMO critics have argued that this means we are not safe from invasion by transgenes that might wreak havoc on our bodies. Yet the empirical fact remains that we have been eating DNA throughout all of human history, and our genomes do not show evidence of being compromised or contaminated by the genes of any of the countless types of foods that humans have eaten across the ages.

Still, eating *any* food, GMO or not, carries a small but finite element of risk. Some degree of risk attaches to practically any human activity. That is why, when scientific bodies endorse GMO foods, they do not claim that they are categorically or absolutely safe but only that GMO foods currently marketed are at least as safe as traditional foods. This view has been expressed repeatedly over the years, in one form or another, by countless scientists and scientific organizations that have studied GM foods. In 2003, for example, a Society of Toxicology position paper stated, "The available evidence indicates that the potential adverse health effects arising from biotechnology-derived foods are not different in nature from those created by conventional breeding practices for plant, animal, or microbial enhancement."

More than a decade later, in 2016, the US National Academy of Sciences published the results of a large-scale study of genetically engineered crops. The scientists conducting the study did a literature survey of more than 800 published papers, heard from 80 speakers at three public meetings and 15 webinars, and read more than 700 comments from members of the general public. And at the end of it all they reported, "The study committee found no substantial evidence of a difference in risks to human health between currently commercialized genetically engineered (GE) crops and conventionally bred crops."

■ Finally, discussions of GMO food safety almost invariably focus on the potential risks or harms of consuming such foods. Rarely has there been a consideration of the actual, demonstrated, and genuine health *benefits* of genetically modified crops. But such benefits do exist and should be recognized. Some of the benefits are immediate and obvious; others are more obscure and indirect.

Two of the most common alterations made to genetically modified crops are those that confer insect resistance and herbicide resistance in plants. In 2015, such crops were being grown on about 12 percent of the world's planted cropland, representing a total area of 444 million acres. There were herbicide-resistant varieties of maize, soybeans, cotton, canola, sugar beets, and alfalfa, and insect-resistant varieties of maize, cotton, poplar, and eggplant. Herbicide-resistant crops resulted in both greater yield and a decline in herbicide use. The planting of insect-resistant crops resulted in the reduced use of insecticides, which not only lowered costs but also lowered the rates of insecticide poisonings and deaths among the ranks of farmers who formerly sprayed synthetic insecticides, a hazard that no longer exists when the plants themselves do the job of repelling insects.

The introduction of *Bt* cotton to India provides an illustration of these effects. Several varieties of *Bt* cotton were genetically modified (principally by Monsanto) by the addition of genes from the soil bacterium *Bacillus thuringiensis*, the "natural pesticide." The first plantings of *Bt* cotton were conducted in India in 2002 to control the bollworm and other insect pests that were endemic to the country. Over the next three growing seasons, insecticide use declined by an average of 41 percent. Since the insecticides that had been used previously were highly toxic synthetic chemicals, these reductions produced positive consequences for the health of the farmers as well as for the environment.

A second positive effect of *Bt* cotton in India was an increase in crop yields by an average 37 percent across the same three growing seasons, as a result of more effective pest control and lower crop losses. Reduced insecticide costs coupled with higher crop yields translated into profit increases of 89 percent. Because most of these gains were to smallholder farmers in poor, rural areas where picking cotton was primarily a female activity, this rise in income meant better access to food, with positive effects on women's income and improvements in child nutrition and health.

These health benefits were derived from GMO cotton that was insect resistant. Additional benefits accrued from GMO crops that were herbicide resistant. Prior to the planting of such crops, farmers controlled weeds by burning them and by tilling the soil. Burning contributes to air pollution, while plowing causes soil erosion and the release of soil carbon into the air as CO_2, a greenhouse gas, while increasing fossil fuel consumption by tractors. Crops that were herbicide resistant led to decreases in all of these factors, with corresponding positive effects for human health and for the environment.

That these effects are indirect, and in some cases unintended, does not make them any less real. They are true and genuine positive health benefits derived from the planting, sale, and use of genetically modified crops.

■ This, then, is the situation in short. Humans have been eating genetically modified foods of many types for more than two decades with no known ill effects. The comparison of epidemiological datasets between nations where GMOs are consumed and where they are not shows no difference in disease levels or types of health problems.

Anti-GMO biases are attributable more to perception than to facts. And those perceptions are products of multiple influences including the existence of stringent government regulations on all aspects of GMO research and development, bans of GMOs by some countries, activist fear and propaganda campaigns, cognitive miserliness, anti-GMO virtue signaling, mass destruction of field trials, prejudicial words and images, and anti-GMO memes that persist despite abundant contrary evidence.

Against those influences, negative perceptions, and preconceptions stands the fact that across more than two decades the scientific academies and study groups of many nations—including even in those European countries where GMOs are in fact banned—have concluded that genetically modified foods are no less safe than conventional foods.

The empirical safety record of GMO food consumption is explained by the facts that (1) the genes inserted into a fruit or vegetable genome are minuscule as compared to the size of the genome itself; (2) the inserted genes are well characterized and known; (3) any unintended effects brought about by gene insertion are removed by DNA repair mechanisms and by repeated backcrossings; (4) before being released to the marketplace, GMOs must survive many rounds of government-mandated safety reviews; and (5) recombinant DNA, like ordinary DNA, is largely degraded in the digestive tract, rendering impotent the gene's alleged ability to cause harm.

All of these factors combined amply support the conclusion that genetically modified foods are at least as safe to eat as traditional foods. And, when added to their proven health *benefits*, there is a way in which GMO foods are, on balance, even *safer* than conventional foods.

For GMO opponents, an inconvenient truth.

7
GOLDEN RICE 2

When Ingo Potrykus, Peter Beyer, and Adrian Dubock traveled to the Philippines in January 2001, they gave the packet of Golden Rice seeds and the six tubes of genes to Gurdev Khush. Among the world's rice breeders, Khush was a legendary figure with a reputation on par with that of Peter Jennings, who had first come up with the idea for provitamin A rice back in 1984. Khush had been the head of plant breeding at the International Rice Research Institute since 1972, and during that time had developed more than 300 new rice varieties. One was a type called IR36, which according to IRRI was "the most widely planted variety of rice, or of any other food crop, the world has ever known." It was a semidwarf rice, so called not because its grains were tiny but because the plants themselves were relatively short, which allowed them to produce heavier seed heads without falling over, or "lodging," as rice farmers called it. IR36 was also naturally resistant to a good number of pests and diseases, and matured more rapidly than other varieties. For all these reasons it became a favorite among rice growers.

In 1999, when Golden Rice was first announced, Khush had predicted that it would be available to farmers in just three years, by 2002. That might have been true if Golden Rice had not been genetically engineered. "None of us knew at the time what genetic modification meant for a breeding program," Potrykus said much later. If the product had been an ordinary, conventionally bred rice variety, then transferring the desired trait, the presence of beta carotene in the kernels, into an established rice variety that was suitable for growing in the tropics would have required eight backcross generations, or about three years' worth of crossings. But the fact that it was a GMO meant that Golden Rice would face so many different kinds of obstacles that those initial three years would stretch out almost endlessly, into a phenomenon resembling the ancient Greek philosopher Zeno's Race Course Paradox, in which the finish line recedes farther into the distance as the runner approaches it.

Gurdev Khush would have been the ideal breeder to deliver a usable Golden Rice product within three years, but as it turned out he was due to retire in 2002. And so after getting the backcrossing work started he handed over the task to his successor at IRRI, Parminder Virk.

Virk was comparatively new to the institute, but he was a senior scientist in plant breeding and was fully equipped for the job. The proof-of-concept Golden Rice was of a japonica rice variety, Taipei 309. Japonica is well suited for lab transformation experiments but doesn't grow well in South and Southeast Asia and is not much liked by consumers. The beta carotene trait therefore had to be transferred into indica rice, of which there are several discrete local varieties. Virk's goal was to use conventional backcrossing techniques to transfer that trait into a few different indica types.

Backcrossing is a standard method used in conventional plant breeding programs. The point is to breed a desired trait into a plant that does not already contain it. In conventional backcrossing, the donor variety possessing the desired new trait is crossed with the receiving variety multiple times, to make an offspring that in the end is almost exactly the same as the receiving variety except for the addition of the new attribute.

"After six generations of repetitive backcrossing and visual selection," Virk said, "the plants will have nearly 99 percent of the original genome reconstruction, but still slightly more than one percent from the donor, including the desired new trait."

He worked with four indica rice types: the popular IR36 rice that had been developed by Gurdev Khush himself, plus IR64, which is eaten all over Asia, together with a high-yield variety that is popular in Bangladesh (BR29), and another that is a particular favorite in the Philippines (PSB Rc82).

Separately, Potrykus, Beyer, T. T. Cuc Hoa of the Cuu Long Delta Rice Research Institute in Vietnam, and other members of the ETH team also introgressed the Golden Rice gene into the IR64 rice line and into another indica variety, MTL250, which was widely used in the Mekong Delta. This effort had produced the second version of Golden Rice, GR 0.5. In the process the researchers cleaned up and streamlined the DNA construct in an attempt to make the GMO rice, as they said, more "amenable to deregulation."

Deregulation, as we have seen, is the process that ultimately leads to the registration of a GMO product and provides for its commercial or noncommercial use by consumers. Deregulating Golden Rice was a necessary step in getting it to the people who could benefit from it, and it was a struggle that would engage the inventors and their colleagues for years into the future. To expedite the process, to distribute the workload, and to get the beta carotene gene introduced into locally popular rice varieties in different countries, Potrykus also provided seeds to rice research institutes in India and Bangladesh.

But these were only the initial steps of a much larger international effort to get the rice to scientists, farmers, and ultimately to the vitamin A–deficient poor in as many countries as needed it. To accomplish these goals, Potrykus, Beyer, and Dubock created an infrastructure to oversee and advance the future progress of Golden Rice. First they established the Golden Rice Humanitarian Board, based in Zurich. The board consisted of about a dozen unpaid volunteers who were experts in rice science. They included the two coinventors and Dubock, along with Gary Tonniessen, Gurdev Khush, and representatives from various rice research institutes and universities in Europe and the United States. Together they would help guide Golden Rice through its development phase, and in particular they would advise researchers and rice breeders about how to navigate the regulatory maze that was to face them when the Cartagena Protocol came into effect in 2003.

The second part of the infrastructure is the Golden Rice Network, based at IRRI and headed by Gerard Barry, the institute's Golden Rice project leader. Its purpose is to facilitate the hands-on processes of seed transfer, breeding, and finally the distribution of the successfully bred rice lines to smallholder farmers of Asia. The network consists of rice research institutes, breeding stations, and universities in the Philippines, India, Vietnam, Bangladesh, and China. The hope was that with this alliance of people and institutes collectively working toward a common goal, Golden Rice would be in farmers' fields within a very few years—if not everywhere, then at least somewhere.

The first meeting of the Humanitarian Board was held at Zeneca offices in Fernhurst, United Kingdom, in August 2000. There the group was immediately confronted with a key strategic decision, one whose consequences would reverberate long into the future. Had Golden Rice been a product of conventional rice breeding, the process of getting it approved for release would be comparatively straightforward. But the fact that the rice was a genetically modified organism effectively propelled it into another breeding and regulatory universe.

A crop that is a product of traditional breeding is field-tested to grow hundreds to thousands of specimens in real-world conditions, during which time the plants are subjected to all the stresses and strains that affect crops: weather variations (drought, heat, etc.), plus pests and disease. From those field-trial specimens, which are evaluated for yield, taste, hardiness, and other qualities, the breeder selects one that will breed true and produce a variety that is genetically distinct, that is stable across time, and that when planted produces crops of a uniform appearance. It might take several years of field trials to achieve the ideal result, but when the most favorable variety is identified, that specific type is the one presented to regulatory officials for approval and registration and for ultimate release to the public.

Time consuming though it already is, that process runs into a catch-22 when the crop in question is a genetically modified variety. Such a new plant type is technically called a "transformation event." In the occasionally baffling world of scientific nomenclature, that language is unfortunate for several reasons. Originally, the term referred to the successful insertion of a transgene into a chromosome. That is, it referred to a *process*, which is after all what an "event" is, by definition. But the term's meaning has since been extended to include not only the process of transgene insertion but also the *result* thereof, in the form of one or more identical clones of a genetically engineered crop variety. In other words, "event" now also means an individual object or string of objects, a genetically modified plant line.

That this is linguistically impossible and conceptually bizarre was the least of the term's problems. The term connotes something momentous—an event!—like a moon launch or Hollywood movie premiere, whereas what it refers to in this case is really just a new plant type, albeit one that is a product of genetic engineering. Moreover, the term adds an aura of foreignness, of scientific doubletalk, to the subject of genetic modification (already arcane enough), which is off putting to many and overly weighted down with a superabundance of unfamiliarity, strangeness, and negativity. As a piece of rhetoric, a "transformation event" is in perfect alignment with the foreboding monster that Dr. Frankenstein created in his lab, which had been a "transformation event" in the original sense.

But mere nomenclature was not the catch-22 associated with getting a GMO crop line approved. The catch-22 was the requirement that any new transformation event—any new GMO variety—has to receive official government approval *before* it is released into the environment, that is, before field trials on it can begin. But how can you identify which plant line to present for government deregulation and approval when examples of the crop type haven't yet been subjected to real-world growing conditions? You have to do so based on molecular data pertaining to the different transformation events, along with phenotypic observations of the plants that have been grown in greenhouses. But greenhouses are highly artificial, controlled environments, with conditioned air, supplemental lighting, constant relative humidity, small temperature variations, and a strictly regulated day-night cycle, making the plants nurtured inside them essentially laboratory specimens.

To make matters worse, every individual GMO transformation event that a breeder puts forward has to gain separate approval for every country it is intended for, and each approval is a long and costly process—to the tune of several million dollars—because of all the mandatory allergenicity, toxicity, and safety testing, and the molecular data characterizations required, which are comparable to what is needed to gain approval for a new drug. It would be more efficient, therefore, as well as more cost effective, to seek approval for only *one* specific transformation event chosen on the basis of molecular information and greenhouse tests. But doing so is a gamble, for the variety chosen on this basis might not perform nearly as well in the real world as it had in the lab.

Nevertheless, when the Golden Rice Humanitarian Board met in August 2000, its members were fully cognizant of the upcoming regulatory hurdles that would have to be overcome when the Cartagena Protocol came into effect, and so, for better or for worse, they decided on a strategy of *one transformation event everywhere.* This meant that they would have to select one specific parent Golden Rice type that would be used forever afterward as the archetype or source material for transferring the beta carotene trait into the many different local varieties in all the several countries for which it was intended. This specially chosen entity would be called the "lead event."

"All variety development worldwide must be based on one lead event," Ingo Potrykus said later. "This strategy has the advantage that one set of regulatory data is valid for the regulation of all varieties in most national regulatory systems in place for GMO crops, which has potentially important cost reduction implications."

Potrykus was well aware of the chance he was taking with this approach. "The new variety must be effective in the field," he said. "None of the tons of molecular data tell you anything about that. In practice you can have invested decades to fish out the putative lead event with perfect molecular data, but in the field, it fails."

Which in fact is exactly what would happen.

■ As is true of practically any new technology, improvements and refinements come with time. And so too it was with Golden Rice.

It is a general rule in biotechnology that one of the most common fates of new genetically engineered plants is that the targeted, desired trait is not expressed strongly enough in the prototype. In the case of the proof-of-concept Golden Rice, the trait in question, the buildup of beta carotene in the kernels, was indeed expressed only weakly. The amount of total carotenoids (reddish pigments) formed in the prototype version was 1.6 micrograms per gram (1.6 µg/g) of dry, uncooked rice. A microgram is one millionth of a gram, an amount so tiny as to be invisible to the naked eye; but, then again, vitamins are needed only in incredibly small, generally invisible amounts to begin with. Still, 1.6 micrograms does not represent an especially great quantity of beta carotene, as only a portion of the carotenoid consists of beta carotene (some other components being alpha carotene and ß-cryptoxanthin).

It was the low level of beta carotene expression in this, the prototype version of the rice, that Greenpeace seized upon and lampooned to maximum effect with its claim that "an adult would have to eat at least twelve times the normal intake of 300 grams (3,600 grams, or eight pounds) to get the daily recommended amount of provitamin A." The claim was clearly a bogus factoid inasmuch as the bioavailability of the beta carotene in the rice (the amount that is actually converted to usable vitamin A in the body) had not yet been established. But, ironically, the reason that the inventors did not yet know the bioavailability value was because of the difficulty of growing sufficient quantities—on the order of hundreds of kilograms—of the grain in order to conduct feeding trials when the rice plants were by law confined to small cabinets in biosafety lab greenhouses and were not permitted to be grown in the field, where arbitrarily large amounts could be planted, cultivated, and harvested. This was yet another GMO-specific catch-22 that materially delayed the development and deregulation of Golden Rice.

When Golden Rice critics complain, as they routinely do, that the rice has been in development for 20 years and "there is still no product," their implication is that there is something wrong with the rice itself or the technology that produced it. What's really at fault for a large portion of the 20-year wait, however, are overly restrictive government rules, regulations, and catch-22s.

The original Golden Rice was never intended as a panacea or magic bullet in the first place. As Beyer and Potrykus wrote in 2002, the prototype rice "is not expected to provide 100% of vitamin A in the diet but to add to present intakes to reach vitamin A sufficiency. . . . Efforts are underway to triple the amount of the provitamin in the endosperm at a minimum."

In an attempt to improve beta carotene levels in the endosperm, in 2002 the Golden Rice Humanitarian Board asked licensees of the Golden Rice Network to create about 1,000 new transformation events, from which pool an improved version of the rice could be selected and ultimately put forward as the "one lead event." Their request led to an international effort that in fact resulted in a new version, Golden Rice 1 (also GR 1 and SGR1). Help came from an unlikely source: the Japan Tobacco Company, which had developed a method for removing selectable marker genes entirely. As was true of the earlier version, GR 0.5, removal of an antibiotic marker gene would make the prospective new rice more acceptable to regulators.

As a result of a joint venture between the Japanese and Syngenta scientists, some 800 new transformation events were produced in Japan. Some of these events produced as much as 13 micrograms per gram total carotenoids in the endosperm. An additional 200 events created in Peter Beyer's lab expressed a maximum of 6 micrograms per gram.

This was the third version of Golden Rice, GR 1. It constituted an advance in the technology because of the increased beta carotene levels in the kernels, caused by additional changes in the transgene, including a new endosperm-specific promoter that was part of the genetic construct. But there was yet an even better version in the offing, one that would completely eclipse GR 1, which as a result makes only fleeting appearances in the scientific literature.

The fourth and final version of the rice, Golden Rice 2 (also GR 2 or SGR2), was a product of two teams of Syngenta scientists, one at the company's International Research Center at Jealott's Hill in Berkshire, west of London, and the other in the United States at Syngenta Biotechnology in Research Triangle Park, North Carolina. The overall project leader was Rachel Drake at the Jealott's Hill location. The motivating idea behind this new research effort was the scientists' perception that there was some as yet unknown limiting factor in the rice kernel that was preventing the expression of even higher amounts of carotenoids in the endosperm. Suspicion centered on the PSY protein, phytoene synthase, in the rice kernels, which was a product of the daffodil *psy* gene.

The daffodil *psy* gene had been used not only in the prototype Golden Rice but also in both of the other two versions produced to date (GR 0.5 and GR 1). But there were other *psy* genes coding for different types of PSY proteins in other plants, and so to test the hypothesis that a new phytoene synthase gene might lead to increased carotenoid levels, members of the two groups took *psy* genes from four other plant species—carrot, tomato, bell pepper, and maize—put them into rice plants, and compared the results of the respective transformations.

As luck would have it, the daffodil *psy* gene performed the worst of all. The clear winner was the maize *Zm psy* gene (standing for *Zea mays psy*), which produced accumulations of carotenoids that reached a maximum of 36.7 micrograms per gram in one of the transformation events, a level that was fully 23 times greater than the amount present in the Golden Rice prototype, GR 0.0.

"All *psy* [genes] proved more efficacious than the daffodil *psy* that was used in the original Golden Rice," the team members reported in their 2005 article on the subject in *Nature Biotechnology*. The daffodil PSY protein itself was thus revealed to be the limiting factor in carotenoid accumulation. In effect, it was a barrier to higher expression levels of carotenoids, but nobody knew that at the time that the original Golden Rice was invented.

The discovery that the *Zm psy* gene yielded such a greatly increased level of carotenoid was a major technical milestone in the development of Golden Rice. It represented an entirely new, fourth-generation version of the plant, GR 2. This was clearly the Golden Rice of the future, a powerful new variant that could potentially save the sight and lives of millions.

Whereas Greenpeace had objected to the original Golden Rice on the grounds that it provided too little beta carotene to make any positive contribution to human health, it now objected to Golden Rice 2 on precisely the opposite ground, that it provided *too much* beta carotene to the consumer. "Large doses of beta-carotene can have negative health effects," the group reported in *Golden Illusion: The Broken Promises of Golden Rice*. While it is a truism that high enough doses of practically any substance can be harmful—the first principle of toxicology being the adage, attributed to Paracelsus, that "the dose makes the poison"— the implication that the amount of beta carotene in Golden Rice poses a threat to human health is unrealistic at best.[*] In general, excess amounts of beta carotene are excreted from the human body without being converted to vitamin A.

But for all its increased levels of provitamin A, SGR2 (as the Syngenta scientists referred to it in print) was still a creature of the laboratory, as were all the individual transformation events that the researchers had experimented with: SGR2A1, SGR2B1, SGR2C1, and so on, 23 individual transformation events altogether.[†] Those 23 transformation events became the pool from which the *one transformation event everywhere* would be drawn.

Some of those events expressed higher levels of carotenoids than others, but the highest-yielding event would not automatically be the best candidate for the lead transformation event. It might be undesirable for other reasons: perhaps it was slow to mature; or produced fewer seeds; or was less disease, drought, or pest resistant. The situation was much like a murder mystery in which 23 suspects, including the murderer, were gathered in a room, and the task was to find out who the culprit was, but without enough information to implicate the guilty party uniquely. In the case of the 23 transformation events, likewise, there was a paucity of actual clues, because all the events were still just prisoners of the greenhouse.

"It is, however, the carotenoid content achieved in Golden Rice 2 plants under field conditions, when the transgenes have been introduced

---

[*] Studies showed that megadose-level (30+ mg/day) intakes of beta carotene were associated with negative effects with respect to cancer prevention among smokers and those who had been exposed to asbestos. But these were "unusually high doses that greatly exceed normal dietary levels" and could not be attained by consuming the amounts of beta carotene provided by GR 2.

[†] The total number of primary transformants produced by the Syngenta researchers was 619, out of which they retained for evaluation only a much smaller group of transformation events.

by backcrossing into locally adapted varieties that is the ultimate determining factor in their contribution to the alleviation of vitamin A deficiency," the Syngenta scientists wrote. "Further research and development activities are required before these events could be released from regulation."

■ The SGR2 experiments ended in 2004, which was to be a pivotal year in the development of Golden Rice. The least of it was that the United Nations had declared 2004 the International Year of Rice. This was essentially a symbolic gesture, but it called attention to the facts that rice was the dietary staple of more than half the world's population and was the world's most important food.

Far more consequential was that, in October of the same year, Syngenta, to mark World Food Day, October 16, announced that it was abandoning all commercial interest in Golden Rice. Although the company gave no reason for this action, Syngenta had apparently decided that the rice was not really a commercially viable product. After all, beyond its possible value as a health food or novelty item, it had no particular sales advantage in developed countries, where vitamin A deficiency is not a problem, whereas in developing countries it would be given to farmers at little or no cost. So there was no realistic hope of deriving a reasonable profit by commercializing it.

With its abandonment of its commercial interest, Syngenta removed from the discussion one of the most common accusations that GMO opponents leveled against Golden Rice, namely that the company was supporting the product for profit rather than for humanitarian reasons. That objection no longer had any plausibility or foundation in fact, although anti-GMO activists would tirelessly persist in making the claim anyway.

Simultaneously, Syngenta also announced that it was donating all rights to the product, including all Golden Rice seeds, genes, plant lines, scientific results, and related technologies, including all of the SGR2 transformation events, to the Golden Rice Humanitarian Board. Finally, the company reaffirmed its commitment to continue supporting further research and development, including funding future field trials.

When field trials finally began, also in 2004, they were not conducted in the Philippines, India, or in any other Asian country, but rather in the United States, on a small plot of farmland just outside of Crowley, Louisiana. Rice is the major crop in the area. In fact, it is such a fixture of the scene that the town hosts a yearly International Rice Festival, bills itself the "Rice Capital of America," and sports the official town motto, "Where life is rice and easy."

But none of those things explained why the world's first open-air, real-life field trials of Golden Rice were held here. There were two main reasons, the first of which was that none of the target countries for Golden Rice—the developing countries of South and Southeast Asia—yet had biosafety regulations in place for handling genetically modified organisms, whereas Syngenta had attached to the agreement with Golden Rice licensees (i.e., the research and breeding institutions that made up the Golden Rice Network) the condition that no field trials could occur in any country that lacked a national regulatory framework covering GMOs.

Adrian Dubock explained in 2017 that it had been necessary to attach that condition to the agreement "as the regulations were anticipated but had not yet come into force." Nobody involved with Golden Rice wanted to risk beginning a field trial only to have it outlawed once it was under way.

Paradoxically, in view of all the trouble that GMO regulations were to cause once they existed, in this case it was the *lack* of a regulatory framework, rather than an excess of regulation, that prevented field trials from being held in countries where vitamin A deficiency was a problem. The United States, by contrast, *did* have a regulatory structure in place (principally, the FDA's 1992 Statement of Policy: Foods Derived from New Plant Varieties), but because the United States was not a signatory to the Cartagena Protocol, its regulations were not hamstrung by the Precautionary Principle and its cargo of worst-case expectations and scenarios, and instead were relatively benign and reasonable.* Even so, it

---

* As reflected, for example, in the FDA's statement that "the agency is not aware of any information showing that foods derived by these new methods differ from other foods in any meaningful or uniform way, or that, as a class, foods developed by the new techniques present any different or greater safety concern than foods developed by traditional plant breeding."

took six months for Ingo Potrykus to get government permission to transfer his rice to the United States.

The second reason for holding the field trials at Crowley was that it was the home of Louisiana State University's Rice Research Station. It was the nation's first rice experimentation center, established in 1909, and it was a logical place to conduct any type of rice research, golden or not.

"When possible, I would prefer to have these field trials at this station," Peter Beyer remarked on a visit there in 2005.

Field trials with Golden Rice 1 started at Crowley in time for the summer growing season in 2004. For this first open-air test, the beta-carotene-inducing transgene had been bred into an American type of long-grain rice, Cocodrie, a japonica variety that had been developed at Crowley during the 1990s and grew well in Louisiana. Over the years, Cocodrie had become the most widely grown rice variety in the United States.

In June 2004, the station's resident coordinator, Steven D. Linscombe, together with his staff of breeders, planted the transgenic Cocodrie seeds in an open field (Field 12), which was bordered by a rural highway on one side and on another by stands of control groups of rice plants of the same variety. This was the very first time that any type of Golden Rice had actually been exposed to the full light of day, anywhere in the world.

The rice was ready for harvest in September 2004, and for the grand occasion several members of the Golden Rice Humanitarian Board were in attendance, not only Potrykus, Beyer, and Dubock, but also Gerard Barry of IRRI, Rob Bertram of USAID, and Jacqueline Paine of the Syngenta Jealott's Hill lab in Berkshire. There is a picture of all of them lined up in a row, each standing behind a pair of rubber boots they were about to don in order to plod through the lines of plants.

The test results were both surprising and not. The surprise was that the field-grown rice accumulated about four times more beta carotene than the plants that had been raised in the greenhouse, probably due to better growing conditions in real-world soil, sky, and sunlight. The field-grown Cocodrie plants expressed an average of 6 micrograms per gram of total carotenoids, but in some plants the level was as high as 8 micrograms per gram. The unsurprising result was that, otherwise, the genetically engineered plants differed not at all from the non-GMO control plants in terms of properties such as plant height, days to flowering and seed set, 100-seed weight, total biomass, and other relevant indicators.

The next year, 2005, saw the planting and harvesting of the new Golden Rice 2, also at Crowley. In this case the maize beta carotene gene had been introduced into Kaybonnet, which was another southern US adapted cultivar. At the end of that summer Peter Beyer came back to Louisiana to collect mature rice samples.

"This second trial was not overly successful, because hurricane Catharina [Katrina] hit us just at harvest time," Beyer recalls. "It did not do so much to the field but air transportation was not possible and refrigeration failed because of many brownouts. However the trial did help designing the follow up studies done in my lab and in the Philippines."[*]

Nevertheless, the results obtained at Crowley established definitively that it was SGR2, and not SGR1, that would be the most successful in farmers' fields and in combating VAD. A separate batch of the rice was sent to Tufts University in Boston, where scientists would conduct the first nutritional and bioavailability studies of Golden Rice.

Beyer, for his part, was confident that the rice would do well. "I'm not worried about the science," he said at the time. "I'm worried about the bureaucracy."

Potrykus, too, was worried about the bureaucracy but also about Greenpeace. On March 18, 2005, he had given a public lecture at the Institute for Plant Science in Zurich during which he announced the new Golden Rice 2 and its greatly increased beta carotene levels, reported on the Louisiana field tests of SGR1, and severely criticized government overregulation for retarding the pace of Golden Rice development. In Europe, anyway, as opposed to the developing countries of Asia, strict GMO regulations based on the Cartagena Protocol, and reflecting the restrictiveness of the Precautionary Principle, had arrived.

---

[*] Peter Beyer, email to Ed Regis, April 10, 2018.

"Regulation is set to look at risks, not benefits," Potrykus had said in his talk. "GMO regulation, so far, prevents use of this technology by the farmer."

Three days later, Greenpeace issued a press release focused on Potrykus's talk. Under the headline "Renewed 'Golden' Rice Hype Is Propaganda for Genetic Engineering Industry," the press release said, "Syngenta claims to be trying to help people suffering from vitamin A deficiency, but Potrykus' attacks made it clear that the GE industry uses 'Golden' rice to campaign for an easier marketing of GE seeds. It looks like the industry is campaigning for their GE seeds while abusing people that suffer from severe malnutrition."

Abusing people? None of this wild rhetoric was even remotely true or made any sense, but it showed that Potrykus had every right to be worried. Or at least exasperated beyond measure.

■ Although the Louisiana field trials were a milestone in the development of Golden Rice, demonstrating that it performed better in real-world, outdoor environments than it did in greenhouses and produced useful levels of carotenoids, the trials nevertheless had a considerable drawback from the standpoint of actually getting the biofortified rice to those who needed it. The problem was that neither of the two varieties tested at Crowley—Cocodrie and Kaybonnet, both of which were japonica types of rice—were suitable for growing in the tropical climates and soils of South and Southeast Asia, where farmers grew and people ate indica varieties. That in turn meant that the field trials were also of little help in identifying the "one lead transformation event" that would be used for introgressing the beta carotene trait into the different, locally grown indica rice types in the countries for which it was intended.

And so, for the purpose of determining what the lead event would be, the whole process of putting the trait into locally adapted varieties and then subjecting them to field trials would have to be repeated all over again in the tropics. After some discussion, members of the Humanitarian Board decided that six different Syngenta SGR2 transformation events would be most appropriate for those projects. And by means of a Materials Transfer Agreement—a contract governing the transfer of materials between organizations for use in research as opposed to commerce—they sent the six varieties to IRRI in the Philippines and also to the Indian Agriculture Research Institute (IARI) in New Delhi.

At that point, all previous materials used by the scientific members of the Golden Rice Network had to be destroyed. "An extremely tough decision for our partners," Ingo Potrykus recalled, "but necessitated by regulations governing GMO crops."

The purpose of the regulation in question is to minimize the risk that seeds of those strains previously experimented with but subsequently rejected for further development work would somehow make their way onto farmlands and contaminate crops, much as the genetically modified StarLink corn had gotten mixed in with conventional corn in 1998–2000. The risk of that happening with Golden Rice was low in view of the one thing about it that everybody, both supporters and critics, could agree upon, which is that its kernels, when milled and polished, are visibly different, obviously, almost ostentatiously, yellow. The polished rice is in effect color-coded in a way that makes it easy to avoid commingling its grains with any other rice variety.

That risk was reduced even further by the physical conditions under which the introgressing work in the Philippines and in India was to be conducted, for, in both countries, the backcrossing was by law confined to highly restricted and controlled indoor environments. At IRRI, the work was done in what amounted to a controlled environment biosafety lab. In India, the containment lab was just as cut off from the outside world: it was the National Phytotron Facility located at IARI, in New Delhi. Much like its counterpart in the Philippines, breeders entered and exited the lab quarters by means of an airlock, as if on a space station, and were "sterilized" by high-pressure blasts of air to remove any contaminants that could affect the plant growth experiments going on within or to bring any GMO plant contaminants out into the world at large.

All of which meant that Golden Rice was once again back in hermetically sealed growth chambers, once again a lab plant.

■ It is difficult to produce enough rice to do rigorous animal safety tests with plants grown in such laboratories. But there were other ways to establish the safety of Golden Rice as a food. One was by analyzing the rice and its ingredients, down to the molecular constituents that had been used to genetically engineer it into existence, and then evaluating the safety of each component separately.

Bruce M. Chassy, a professor in the Department of Food Science and Human Nutrition at the University of Illinois, was a specialist in assessing the safety of transgenic crops used as foods and feeds. After its invention in 2004, Chassy undertook a systematic study of the safety of Golden Rice 2 as a food for human consumption and published the results in two authoritative studies, "Golden Rice 2" (2008) and "Food Safety of Transgenic Rice" (2009).

At the level of the grain itself, Chassy noted the obvious facts that both rice in general as well as the beta carotene in other plants have had long histories of safe use as foods. He noted further that Golden Rice contained a lower level of carotenoids than are commonly found and safely eaten in foods such as sweet potato, carrot, and maize, thus debunking Greenpeace's claim that there were unhealthy levels of beta carotene in GR 2.

As for the inserted genes and their components, they too seemed harmless. The maize *psy* gene that had been introduced into Golden Rice 2 had been isolated from a plant, corn, that itself has a long history of safe use in human nutrition. The rice also contained a mannose marker gene, which in the case of Golden Rice 2 was *pmi*, which coded for the enzyme PMI (phosphomannose-isomerase). The *pmi* gene had also been used in GR 0.5, had previously been utilized in other transgenic crops as well, and had been approved for use in foods by the FDA. In addition, *pmi* was found in strains of bacteria that normally inhabit the human intestinal tract.

Indeed, Chassy found that there was only a single genetic component in Golden Rice 2 that did *not* have a history of safe use in human nutrition, and that was the carotene desaturase gene, *crtI*. However, the *crtI* gene, which had been isolated from the bacterium *E. uredovora*, "is

almost certainly present in plant material consumed by humans and animals," and for that reason is not a cause for concern. In any case, rice is
cooked before consumption, commonly by steaming, and this thermal
processing is enough to denature most genes and proteins, which as we
have seen are then rapidly degraded in the stomach in most cases. From
an analytic standpoint, then, every indication pointed to the conclusion
that Golden Rice was as safe to eat as regular rice.

This conclusion was supported by other safety assessments of GR 2.
In 2006, scientists at the University of Nebraska's Food Allergy Research
and Resource Program undertook an allergenicity study of Golden Rice 2.
Their final report on the project, "Bioinformatic Analysis of Proteins in
Golden Rice 2 to Assess Potential Allergenic Cross-Reactivity," ran to
121 pages of data and text.

The object of the study was to evaluate the potential allergenicity of
the three proteins that are expressed by the genes that had been introduced into GR 2. These are the phytoene synthase protein (PSY) that
had been derived from maize, the carotene desaturase protein (CRTI)
that had been taken from the bacterium *E. uredovora*, and the phosphomannose-isomerase enzyme (PMI), from *E. coli*. The scientists proceeded
by comparing several structural and genomic features of those three
introduced proteins against two protein databases of known allergens:
AllergenOnline and the Entrez Protein Database maintained by the National Center for Biotechnology Information, which is a branch of the
National Institutes of Health. Further, the scientists used three search
algorithms to find possible matches between the Golden Rice proteins
and known allergens. All three searches of the two databases returned
negative results.

"This demonstrates that there is not expected to be any significant
risk of cross-reactivity for those who are allergic to known allergens," the
scientists concluded. "In fact, based on these results it would not be possible to identify allergic individuals who would be at heightened risk."

To anyone who might suggest that there were still some residual risks
left dangling, perhaps making the last-resort "unknown safety risks" objection so often leveled by GMO critics, Bruce Chassy had two answers.

First, "Golden Rice should not become mired in a regulatory review that seeks to evaluate and eliminate every conceivable hazard." The reason is that it is not scientifically possible either to identify or to eliminate every last potential danger.

Second, "any potential risks that may be identified with Golden Rice 2 should be balanced against the potential to ameliorate VAD, and, thereby, reduce the loss of life and clinical symptoms due to the nutritional deficiency. The magnitude of this potential impact will only become known for certain if Golden Rice 2 is adopted."

If and when.

BETTER THAN SPINACH

From the very beginning, there was no more persistent, vocal, and intransigent opponent of Golden Rice than the environmentalist organization known as Greenpeace. Its hostility was part and parcel of its broader and long-standing opposition to genetically engineered foods of all types. The organization used every means at its disposal to stop research, field trials, government approvals, and farmer adoptions of GMO crops. Greenpeace issued endless streams of press releases, position papers, and miscellaneous statements concerning the supposed failings, dangers, and environmental and health risks of genetically modified foods.

In addition to all the rhetoric, Greenpeace mounted court challenges to proposed field trials of genetically engineered crops. In 2012 for example, it had filed a petition in a Philippines court to halt a government-backed field trial of genetically modified eggplant. In 2013, the Philippine Court of Appeals ordered that the field trials be stopped. Later, despite a ruling from the Philippine Food and Drug Administration that genetically modified foods "are as safe as and as nutritious as the food derived from conventional crops" and a 2014 study that showed 96 percent of Filipino farmers were willing to sow genetically modified eggplant seeds—and were even willing to pay double the non-GMO seed price if it meant a substantial reduction in spending on pesticides—the Philippine Supreme Court in 2015 ruled in favor of Greenpeace and ordered a ban on field trials.

When Greenpeace was unsuccessful in getting field trials stopped, the organization resorted to what it called "ecotage"—sabotage in pursuit of ecological goals—which meant that it simply destroyed the crops in question. To cite just one case out of many, in July 1999, the same year the prototype Golden Rice had been developed, a group of 20 Greenpeace members, including Peter Melchett, executive director of Greenpeace UK, was arrested for uprooting a 6-acre planting of genetically modified maize on a privately owned farm in Norfolk, England. After a trial in which the perpetrators openly admitted having laid waste to the crops, they were nevertheless acquitted by a Norfolk court, which further awarded to Greenpeace defense costs in the amount of £100,000.

Greenpeace opposition to Golden Rice, however, was especially persistent, vocal, and extreme, perhaps because it was a GMO that had so much going for it and was progressively, albeit very slowly, moving toward eventual approval and release.

Greenpeace claimed that the prototype Golden Rice would do nothing to alleviate the problem of vitamin A deficiency. When the improved Golden Rice 2 was announced in 2005, Greenpeace called it "a technical failure," said that it was "more hype than substance," and claimed that the plants could crossbreed with other rice strains "to contaminate wild rice forever," despite the fact that the presence of beta carotene in the endosperm gave Golden Rice no selective advantage over other types of rice. This is because both it and all other rice varieties—as well as all other green plants more generally—already contain beta carotene, and the genetic pathway responsible for expressing it, in their leaves, stalks, and other plant tissues.

Throughout the years of its strident denunciation and ridicule of Golden Rice, Greenpeace as an organization only once exhibited any sensitivity to the obvious fact that, in seeking to retard its development, approval, and distribution, the group was at the same time prolonging vitamin A deficiency among those suffering from it. That was back in 2001, when Greenpeace's Benedikt Haerlin admitted that Golden Rice posed a moral dilemma for the organization in light of the fact that it could possibly save lives. But he resolved that dilemma just a week later, now saying that "golden rice has not been ruled out as a target for direct action in the future."

Further, in repeatedly claiming that the only "real solutions" to VAD were "increased food diversity, vitamin supplements, and home gardening," plus the utopian "elimination of poverty," Greenpeace was in effect condemning VAD sufferers to their fates until the organization's preferred solutions were implemented, no matter how unrealistic or far-off in time those solutions were.

But Greenpeace had not always been guilty of such callous disregard for human health and welfare, and in fact it had started off as an organization whose purposes, campaigns, and crusades were essentially humane in motive and tone. The group began in 1970 as the Don't Make a Wave Committee, based in Vancouver, British Columbia. Its purpose was to stop US nuclear tests in Alaska, and the name referred to the seismic, atmospheric, and radioactive waves that an underground thermonuclear bomb blast would generate.

On September 15, 1971, a dozen activists, including Patrick Moore, who was then a grad student in ecology at the University of British Columbia, and Bill Darnell, who would coin the name Greenpeace, set off in a chartered halibut fishing boat, the *Phyllis Cormack*, for Amchitka Island, Alaska, in an effort to halt an upcoming test. After a hair-raising voyage that lasted about a week, the group finally made it to the Aleutians. They failed to stop the test but made it onto the CBS News with Walter Cronkite. From its very start, the group seemed to have a matchless knack for garnering publicity.

The next year a separate Greenpeace contingent in New Zealand sailed another boat, the *Vega*, to the island of Mururoa in French Polynesia in an attempt to stop an atmospheric nuclear test to be conducted there by France. They again failed to halt the test, but the campaign received worldwide press attention.

Later Greenpeace offensives included one in 1975 aimed at saving the whales. The plan was to intercept Russian whaling ships that were operating about 100 miles off the coast of California to prevent the whalers from firing their explosive-tipped harpoons at the immense marine mammals. The Greenpeace members who participated in this adventure showed considerable determination and courage as they lowered small rubber inflatable Zodiac boats from the deck of the *Phyllis Cormack* in choppy seas, drove themselves to the whaling fleet, and then repeatedly crossed back and forth in front of the bows of the whalers, who ignored the protesters and calmly fired harpoons over their heads. These daring exploits were captured on film by other Greenpeace members who, once they got back to port, delivered the film to television stations that were only too happy to broadcast footage of their amazing feats.

"We were welcomed into the city of San Francisco as conquering heroes," Patrick Moore recalled later. "There was an explosion of support from around the world." This was another feather in their cap, and it further burnished the already brightly glowing halo effect that Greenpeace always seemed to radiate.

As wild as it was, the save-the-whales campaign was not the peak activist experience for Patrick Moore. That occurred in 1977, on a follow-on save-the-seals campaign, which was an effort to stop the annual bloody slaughter of baby harp seals by Canadians. For this new escapade a Greenpeace group rented a couple of Bell Jet Ranger helicopters and flew to the kill sites, which were on ice floes where the seals were being clubbed to death for their fur. Brigitte Bardot, the French movie star and animal rights activist, had gotten word of the hunt and wanted to have her picture taken with a cuddly seal pup. She, together with a French television producer, a six-person film crew, and an entourage of 80 European journalists arrived at the site, and after much turmoil Bardot got her baby seal photo opportunity.

"By far the most bizarre scene I have ever been in," Patrick Moore said of this episode, "and I have been in some pretty bizarre scenes."

How could anyone not admire this apparently compassionate, courageous, and at times even heroic band of world-savers?

Patrick Moore was one of the original founding members of Greenpeace. He later became head of Greenpeace Canada and finally a director of Greenpeace International, based in Amsterdam. By 2017 the group would have almost 3 million members, with offices in 40 countries, and an annual budget of €236 million ($277 million).

■ By the time the Louisiana field trials were going on using local Kaybonnet and Cocodrie varieties, other versions of Golden Rice were being introgressed into tropical indica rice types at the International Rice Research Institute in the Philippines and at parallel research institutes in India and Bangladesh. But all those lines were being grown inside greenhouses, which meant that they, too, were still the plant-world equivalent of laboratory specimens and were subject to all the limitations pertaining to the controlled and artificial conditions in such environments. It was not until April 2008 that a third open-air field test of Golden Rice took place. The trial was run with a GR 1 transformation event crossed into the popular Asian indica cultivar, IR64.

That test consisted of 20 lines of GR 1 planted on a square of IRRI farmland at Los Baños in the Philippines. This was a small plot that had been outfitted to comply with the GMO crop regulations that were in place there, meaning that the test area was physically isolated from other rice plants by three successive and nested protective barriers. First there was a belt of maize plants that fully enclosed the plot of Golden Rice, which was meant to function as a pollen trap. Surrounding the maize layer was a short metal fence and then a substantial flat trough of water that provided a further barrier and also served to irrigate the plants within. Finally, around the outer edge of the moat there was a second, taller metal fence.

Whether those protective measures were necessary or really accomplished anything much, the plants themselves flourished, grew, and set seeds normally but otherwise did not generate a lot of new information, or even a lot of rice. Parminder Virk, IRRI's senior scientist in plant breeding, gives the reason: "Unfortunately, one week before harvest, a typhoon destroyed the field, turning the whole exercise into a rehearsal for the future."[*]

In any case, this confined field trial was too small to provide significant amounts of new data. "The 2008 planting was the first open planting in Asia but not a proper 'field trial,'" said Adrian Dubock afterward. "It was only one plot." A "proper" field trial would involve thousands of plants at multiple locations, not just hundreds in one place.

---

[*] Adrian Dubock's claim, in "The Present Status of Golden Rice" (2014), that the rice was harvested "just in time before a powerful cyclone would have destroyed it a day later," is not correct.

By the time of the Philippines test, Golden Rice 2 had been bred into a number of rice varieties and was growing in several countries: in Japan, in South and Southeast Asia, as well as in Europe and the United States. Scientists at Syngenta knew precisely how much total carotenoid was expressed in the kernels of each of their 23 transformation events, and other crop scientists and breeders who developed new varieties based on those events also knew that the inserted genetic pathway reliably caused beta carotene to be produced in the grains.

But, even with this increasing body of experimental data about the rice, some unknowns remained. One was the bioavailability of the beta carotene in the rice, that is, the proportion of the beta carotene consumed by an individual that is actually converted to vitamin A, retinol, in the body. This is important since establishing the conversion ratio would demonstrate whether and to what extent eating Golden Rice is effective in delivering vitamin A to those who are deficient. That information, it was hoped, would also refute Greenpeace claims that GR 1 would provide too little vitamin A to be of any utility to sufferers of VAD, while GR 2 would provide so much as to be actively detrimental to their health.

Determining the bioavailability of micronutrients in plant foods is a complex business, however, as it is influenced by a number of interrelated factors such as the way the food is processed and prepared, differences in individual metabolic rates, the role played by other meal components, and so on. The bioavailability of the beta carotene could be established only by experimental test, and to be useful the test would have to be conducted on humans because animals and humans process beta carotene differently,* and because humans, after all, were the intended recipients.

The first experimental demonstration of the vitamin A equivalence of Golden Rice in humans got under way in 2003, under the design and supervision of Guangwen Tang, of Tufts University in Boston. Tang was born in China, had established a track record of more than 20 years of

---

* With the notable exception of the Mongolian gerbil.

research in human nutrition, is the author or coauthor of 87 scientific papers, and was head of the Carotenoids and Health Laboratory at Tufts. She was assisted in this test program by several Tufts colleagues including Robert M. Russell, a physician who is also a member of the Golden Rice Humanitarian Board, and Michael A. Grusak, a botanist at the Baylor College of Medicine in Houston, Texas, who was to grow the rice that would be fed to the test subjects.

In the simplest terms, it is possible to find the conversion ratio of beta carotene to retinol by having test subjects ingest a known amount of it and then measuring by means of a blood test the quantity of vitamin A it is converted to in their bodies. A problem with this, however, is that there will always be some exogenous retinol in the blood from outside sources such as vitamin pills, other carotene-rich foods (spinach, carrots), and even from beta carotene used as a food coloring. There was an ingenious, although expensive, way around this problem, and that was to chemically label the Golden Rice so that the beta carotene that it contributed to the blood could be experimentally distinguished from the serum carotenes and retinol derived from other sources.

Guangwen Tang circumvented the problem in just this way. Her plan was to grow the Golden Rice plants not in ordinary water but in heavy water, deuterium oxide (D_2O). Deuterium is a stable, nonradioactive, and nontoxic isotope of hydrogen, and traces of it would show up only on the beta carotene and retinol derived from the Golden Rice. Heavy water is notoriously expensive (about a dollar per gram), but except for the volunteer work done by members of the Humanitarian Board, nothing about Golden Rice would ever come cheap.

Growing the rice to be used was its own slow-moving chain of events. Mike Grusak started out at first with some SGR1 seeds that he had received from Peter Beyer. Grusak realized soon enough, however, that because of the small size of his growth chamber, together with the low levels of beta carotene in GR 1, he could not grow enough rice for a feeding trial in a reasonable time. So he waited until Syngenta scientists at Research Triangle Park could send him samples of SGR2, with its much higher carotenoid levels, and he proceeded to work with that.

After germinating the SGR2 seeds, Grusak planted the seedlings in trays suspended over a nutrient solution that did not contain any deuterium. They grew in this way for about three and a half months, at which point they began flowering, a key stage in rice growth. He then grew the plants to maturity in a closed labeling chamber and added heavy water to the nutrient solution. The growth chamber was a closed system so that he could collect the transpired heavy water and reuse it in an attempt to contain costs. In this manner it took him about three years to grow enough Golden Rice 2 for the planned bioconversion studies. Finally, after he had milled and polished the rice, he stored the seeds at −80 degrees Celsius until they were to be shipped to Boston.

This batch of rice that had been grown in million-dollar water was ultimately transported to Guangwen Tang at Tufts, where the grains were cooked and analyzed prior to the clinical study. For the actual experiment, the Tufts researchers recruited five healthy, nonsmoking adults (three women and two men) from the Boston area and put them on a strict food regimen. They were told not to take vitamin A or beta carotene supplements for a month before the start of the experiment and then were fed a known and consistent diet for a week. On day 8, all the subjects were fed 200 grams (7 ounces) of the deuterium-labeled Golden Rice. Given what it had taken to grow the grains, this had to be one of the most expensive rice meals in all of history.

Afterward, Tang and her assistants took blood samples from the subjects and analyzed them by sensitive mass spectrometry. The results showed that an average of approximately 3.8 units of beta carotene had been converted to 1 unit of retinol, giving a conversion ratio of 3.8 to 1. This, said Tang in her 2009 report on the project ("Golden Rice Is an Effective Source of Vitamin A"), was "a very efficient bioconversion of ß-carotene to vitamin A." On that basis, she said, "we project that 100 g [3.5 ounces] uncooked rice provides 500–800 µg retinol. This represents 80–100% of the estimated average requirement (EAR) for men and women and 55–70% of the Recommended Dietary Allowance (RDA, derived from the EAR) for men and women, as set by the US National Academy of Science."

This was the first study of the vitamin A value of Golden Rice in humans, and by any standard it was good news for the inventors, Ingo Potrykus and Peter Beyer. It established that their creation could in fact be a reliable and realistic source of vitamin A in humans, and it definitively invalidated Greenpeace claims to the contrary. It provided enough beta carotene to be useful in alleviating VAD but not so much as to be damaging. In addition, Tang and her associates closely monitored the test subjects for a period after the study, and none exhibited signs of any adverse health effects. "This attests to the probable safety of Golden Rice," she stated.

Still, the study had a number of limitations, the first of which was the small sample size of only five test subjects. This was due to the limited supply of deuterium-labeled Golden Rice: there was just not enough to feed more people. Second was the fact that only a single serving of the rice had been consumed by each subject. Third, since beta carotene and retinol are fat-soluble substances, the experimenters included several sources of fat with each breakfast and lunch served to the participants, including a 10-gram pat of butter, roasted cashews, and corn oil. Whether any, or how much, fat might be available to the malnourished, vitamin A–deficient children for whom Golden Rice was intended was an open question. (Although fat is known to help in the absorption of vitamin A, copious amounts are not necessary.) Fourth, the experiment had been conducted with the japonica rice variety Kaybonnet, and laboratory-grown specimens at that, whereas the intended users of the rice would be eating different varieties of real-world indica rice. Still, rice grown in real-world conditions could be expected to outperform a lab version. Finally, the study had been confined to adults, whereas the primary target of Golden Rice was children—as the authors knew all too well.

The next step, then, was to repeat the experiment with children, preferably from a cohort that was as close as possible in age and ethnic origin to those in the Golden Rice target group.

■ Sometime after the close of the bioavailability study, Guangwen Tang and two of her earlier colleagues, now also working with three other scientists in China, performed a second study that evaluated the effectiveness of Golden Rice 2 on a group of healthy Chinese schoolchildren. The children, students in an elementary school in Hunan Province, were of both sexes and ranged in age from six to eight years.

In addition to establishing the conversion ratio of Golden Rice beta carotene to vitamin A in the children, a further goal of the researchers was to compare that ratio with the vitamin A value of pure beta carotene oil, as well as that of spinach. These comparisons would show how well Golden Rice stacked up to other common sources of vitamin A. For this purpose, the experimenters divided up a total of 68 children into three subgroups, one for each source, by a random selection process.

The researchers drew rice from the same supply of deuterium-labeled Golden Rice 2 that Mike Grusak had prepared for the previous experiment. (He had also grown deuterium-labeled spinach by the same method.) On the test day, the experimenters gave all the children their regular meals but, in addition, 23 of the children also swallowed a 30-milligram capsule of deuterium-labeled beta carotene oil, another 23 consumed a 60-gram (2-ounce) serving of Golden Rice, while a third group of 22 ate a 30-gram (1-ounce) serving of spinach.

The results, derived from serum tests as before, gave the following conversion ratios of the beta carotene to vitamin A for each of three substances: pure beta carotene oil, 2.0 to 1 by weight; Golden Rice, 2.3 to 1; spinach, 7.5 to 1.

These were striking results. They meant that it took 2 units of pure beta carotene oil to be converted to 1 unit of vitamin A, while it took just slightly more, 2.3 units, of Golden Rice beta carotene to be converted to 1 unit of vitamin A. Spinach fared the worst of all, requiring 7.5 units for conversion into a single unit of vitamin A. In other words, Golden Rice was almost equal to pure beta carotene oil in its power to deliver vitamin A to these children, and it was about three times more effective than spinach. And so if pure beta carotene oil could be helpful in alleviating vitamin A deficiency, as even Golden Rice opponents acknowledged, then so could Golden Rice itself.

"Our results indicate that spinach, GR, and ß-carotene in oil capsule can all provide children with vitamin A nutrition," the researchers reported. "Of these 3 provitamin A sources, and at the doses administered, GR was as effective as the pure ß-carotene in oil capsule, and both were much more effective than spinach at contributing to the vitamin A intakes of the children. . . . GR may be as useful as a source of preformed vitamin A [retinol] from vitamin A capsules, eggs, or milk to overcome VAD in rice-consuming populations."

And so Golden Rice, finally and at long last, seemed to have proven itself on every criterion: the necessary genes could be transferred into several different strains of indica rice, the transformed plants generally grew normally, and the beta carotene in a bowlful of cooked Golden Rice was as effective as pure beta carotene capsules in providing vitamin A to human beings, including children, and was much better than spinach in doing so. Evidently, Golden Rice was fully up to the task for which it had been created, and it was now ready, after all these years of preparation and trial, to fulfill the hopes of its creators. At least it would be once it had escaped from the lab and had proven itself in the field.

■ Now that Golden Rice had actually been shown to work, GMO opponents were forced into a corner. Their response was to redouble their efforts against it, staking out positions that ranged from the naïve to the unbelievable in a number of desperate attempts to discredit a proven result, or to minimize its value, or to deflect attention elsewhere. The claim that VAD was best combated not by Golden Rice but by "eliminating poverty," for example, was disingenuous at best. Nobody, least of all Ingo Potrykus or Peter Beyer, was against eliminating poverty; but during the years or decades until that fairly Herculean, utopian, and far-off goal could be realized, Golden Rice represented a technology that was cheap, sustainable, easy to implement quickly on a wide scale, and could avert needless deaths and save people from preventable blindness in the interim.

The "more diverse diet" and similar alternative solutions had been advanced many times over the years by many people. Opponents of Golden Rice have claimed again and again that giving out vitamin A supplements, fortifying existing foods with vitamin A, and teaching people to grow carrots or green, leafy vegetables are more promising ways to fight vitamin A deficiency than Golden Rice. But we have seen that each of those remedies faces significant problems. Supplementation programs are expensive, fail to reach people in remote locations, and require a complex record-keeping apparatus and repeated outside interventions. Rice production, by contrast, tends to be domestic and local, and the product is plentiful and available cheaply even in the world's remotest areas. As for food fortification, a specific fortified food might not be consumed by all members of a given population, whereas rice is practically a universal dietary staple of the at-risk target group. Further, fortification programs are expensive and require infrastructure, industrial capacity, constant maintenance, quality control, and so on. The seed of Golden Rice plants, by contrast, could be planted year after year at no cost and would require no vast infrastructure of any kind or repeated external intervention. It would be a health and nutritional regime that would more or less go of itself.

But it is the suggestion that people grow their own carrots and green, leafy vegetables that is the most naïve, impracticable, and unrealistic suggestion of all. The reality is that many of the poor suffering from vitamin A deficiencies live in large cities where there is a lack of arable land to begin with, and in any case few urban residents own or have access to any amount of farmland anywhere. And even if they do, it would be difficult for them to grow and harvest the relevant foods in large enough quantities to satisfy their nutritional requirements and those of their families, on a regular basis for years at a time. Growing any crop is demanding work and a slow process that is vulnerable to seasonal variations and to adverse weather conditions such as drought, extreme heat, monsoon rains and winds, and so on. Turning entire populations of poorly nourished, often unhealthy and weak city dwellers into strong and hearty farmers who could consistently grow and harvest sufficient perishable vegetables is a pseudo-solution to a genuine problem, a romantic fantasy masquerading as a panacea.

Another activist criticism of the effectiveness of Golden Rice was that it would not work because fats are required for vitamin A absorption, whereas children suffering from VAD have diets that are likely to be low in fat. In a recent (2016) journal article, for example, Glenn Stone and Dominic Glover claimed that Golden Rice would not be effective for the undernourished children for whom it is intended because such children "are virtually certain to have poor diets lacking in fats, which the body needs to absorb vitamin A."

But there is experimental evidence against their claim. In a 2007 study of 116 Filipino schoolchildren, some of whose diets were so poor that there was a high prevalence of subclinical vitamin A deficiency among them, a team of researchers found that carotene-rich plant foods improved both their blood serum vitamin A levels and their stores of vitamin A in the liver, even when their diets included extremely low amounts of dietary fat.

"Regardless of the amount of fat ingested by the Filipino children per meal or the total amount of fat they ingested per day, their mean serum provitamin A carotenoid concentration, total body vitamin A pool sizes, and liver vitamin A concentrations were increased similarly," the authors reported. "Thus the dietary fat requirement for optimal bioavailability and effectiveness of plant carotenoids is minimal."

It is clear, furthermore, that double standards were once again at work. If the claim *were* in fact true that there was not enough fat in the diets of target population groups for them to benefit from the beta carotene in Golden Rice, then that same claim would hold with equal force against *any* alternative proposed source of beta carotene, including that found in the alleged cornucopia of home-grown vegetables issuing forth from the ranks of mythic city gardeners. In fact, the purported lack of dietary fat would even prevent the absorption of beta carotene contained in pills or capsules. There is nothing special about the beta carotene in Golden Rice that makes it uniquely vulnerable to the lack of dietary fat. If that fact were a genuine obstacle to Golden Rice beta carotene, it would likewise be an obstacle to the absorption of beta carotene from *any* other alternative source. But that was never mentioned by those who proposed fortified foods, home-grown vegetables, or beta carotene or vitamin A supplements as better solutions to vitamin A deficiency than Golden Rice.

Finally, several critics have argued that members of the target populations would be dissuaded from eating Golden Rice because they regard yellow rice as a sign of mold, dirtiness, or decay or because they
have a cultural preference in favor of white rice. However, an informal
poll in the Philippines showed that people there would not reject yellow-
colored rice if they knew its potential benefits. Further, parents might
feed it to their children even if the parents themselves preferred to eat
white rice. After all, parents in every developed and developing country
permit some highly unusual, even unpleasant, things be done to their
children in the name of health: they let their children's skin be pricked
with smallpox vaccination needles and polio shots, they let doctors pour
vials of oily yellow liquid (high-dose vitamin A capsules) down their
children's throats, and so on. Eating a yellow-colored rice is arguably far
less intrusive than such medical interventions.

Cultural food preferences are in any case acquired tastes, not inherent traits unique to Asians: historically, Asians had long consumed brown,
unpolished rice in great quantities. Rice became "white" only after rice
producers and traders removed its outer layers by milling and polishing
so that the grains weighed less for shipping and could be stored for longer periods without spoiling.

An acquired taste can be unacquired through education or by a process of acculturation. Nutritional scientist Bruce Chassy has argued that
"the real problem with the claim that people will not accept colored rice
is that the critics are also simply ignoring the fact that colored rice foods
are widely consumed around the world. . . . Saffron, annatto or achiote
and turmeric are all extensively used in various countries to produce
golden yellow rice dishes. Black rice was considered so desirable in ancient China that only the emperor was allowed to consume it. The Bhu-
tanese prepare red rice, and blue rice is a specialty in Malaysia."

Blue rice?

"Perhaps the best way to test if GR is acceptable to consumers," Chassy
said, "is to allow consumers the choice of deciding whether they want to
plant it and/or grow it for themselves and their children." The choice for
many, after all, is between white rice and healthy kids.

■ Shortly after the founding of Greenpeace, a Canadian journalist by the name of Bob Hunter told Patrick Moore something that he later regarded as prophetic.

"Pat, this is the beginning of something really important and very powerful," Hunter said. "But there is a very good chance it will become a kind of ecofascism. Not everyone can get a Ph.D. in ecology [which Patrick Moore had]. So the only way to change the behavior of the masses is to create a popular mythology, a religion of the environment where people simply have faith in the gurus."

By the early 1980s, Moore decided that Greenpeace had indeed crossed the line from its original pursuit of a benign and humane environmentalism to a kind of anti-scientific radical fundamentalism. The group had turned away from preventing things that were genuinely destructive—nuclear tests, whaling, seal killing—to opposing things that were actively helpful to humanity. Chlorinated drinking water, for example.

Greenpeace's opposition to chlorine stems from the fact that the element is toxic. Chlorine gas, after all, is lethal and was used as a chemical weapon during World War I. It killed people then and still can. And so, to be safe, Greenpeace wants to impose a worldwide ban on chlorine in all of its forms, including even its use in PVC plumbing pipes and in drinking water supplies.

"For me," Moore remarked later, "this was when what had been science-based policy turned into a kind of religion based on belief rather than facts or evidence, as Bob Hunter had predicted years before."

Moore left the organization in 1986 and turned his attention to aquaculture (which Greenpeace also opposed), raising farmed salmon in Winter Harbour, British Columbia, where he lived. Salmon farming, he believed, took pressure off overfished wild stocks so that their dwindling populations could be replenished.

Greenpeace, meanwhile, continued its opposition to arguably benign, albeit controversial, technologies including nuclear power and, of course, GMOs, in particular Golden Rice.

A month after the publication of Guangwen Tang's Hunan study results in 2012, Greenpeace East Asia issued a statement that condemned the use of Chinese children as "guinea pigs," dismissed the rice as "a potentially dangerous product" (citing no evidence), and said that the experiment was "essentially yet another example of big business hustling in on one of the world's most sacred things: our food supply." But by this time big business had been out of the Golden Rice picture for eight years. Back in 2004 Syngenta had donated its rights to Golden Rice technology to the nonprofit Golden Rice Humanitarian Board and had summarily renounced all commercial interest in the product. Tang, likewise, had not been financed by "big business" but by government grants.

But Tang's experiment, with its scientific proof that Golden Rice was effective in supplying provitamin A to children, ran counter to Greenpeace's official doctrine (or dogma) that the rice was "potentially dangerous," and so her research had to be discredited or invalidated somehow. And indeed, as it turned out, the Greenpeace East Asia press release precipitated a number of investigations, in both China and in the United States, that in the end spelled big trouble for Guangwen Tang and for her 2012 paper, "ß-Carotene in Golden Rice Is as Good as ß-Carotene in Oil at Providing Vitamin A to Children."

It is sad but true that the historical narrative of Golden Rice development is marked by one big scandal and one big mistake. Guangwen Tang's experiment on the Chinese children was the locus of the scandal, whose aftereffects persist even today.

Tang had designed the study by 2003 and in that year had received support for it from the US National Institutes of Health in the amount of $306,160. By the time she actually conducted the experiment, which was in July 2008, additional monies had come from the US Department of Agriculture and from the Chinese National Technology Research and Development Program. The study had been approved by the Tufts Institutional Review Board (IRB) in 2003, and again in 2008, as well as by the Ethics Review Committee of the Zhejiang Academy of Medical Sciences in 2003. And so, from all outward indications, the relevant rules, regulations, and procedures covering the experiment had been complied with . . . except for what Tang and her research collaborators in China had actually told the parents in 2008 about the precise nature of the rice to be ingested by the experimental subjects, their children.

After the Greenpeace condemnation of the study as "gambling with the health of these kids," Tufts University launched three separate investigations into what exactly had happened in China. Tufts inaugurated both internal and external reviews of how Tang's study had been conducted, as well as a separate "scientific review to determine whether the journal manuscript accurately reported the study research methods, measurements and findings."

Meanwhile, the Chinese Center for Disease Control undertook an investigation of its own. It found that the experimenters had not in fact disclosed to the parents that the rice had been of a genetically modified variety. The journal *Science* published a photograph of an English-language version of the original informed consent form that the researchers had disclosed to the parents. It was dated May 6, 2008, and stated that "Golden Rice is a new rice which makes ß-carotene, thus given [*sic*] the rice a yellow (Golden) color." Further, according to the Chinese CDC, the parents were shown only the signature page of the six-page consent form, which also failed to mention that the rice was a GMO. In consequence of these findings, the CDC fired the three China-based researchers who had taken part in the project and offered financial compensation (in the amount of $12,800) to each set of parents of the children who had been fed Golden Rice.

The Tufts investigations took almost a year to complete. But on September 17, 2013, the university issued a statement in which it essentially confirmed what the Chinese CDC had discovered and added several adverse findings of its own, including the charge that Tang did not fully comply with the Tufts IRB requirements and failed to obtain IRB approval for the changes she had made to the study protocol prior to implementing them. In light of these and other lapses, the university prohibited Tang from conducting human subjects research for two years, "during which time she will be retrained on human subjects research regulations and policies." The university also imposed miscellaneous other penalties and restrictions on Guangwen Tang.

In the wake of all this censure, Guangwen Tang closed her lab—the Carotenoids and Health Laboratory at Tufts—in March 2014 and on April 1, 2014, formally retired from the university.

Still, there was a silver lining to all this, in the fact that the Tufts statement completely vindicated the strictly scientific elements of the experiment, concluding that the university's "multiple reviews found no concerns related to the integrity of the study data, the accuracy of the research results or the safety of the research subjects. In fact, the study indicated that a single serving of the test product, Golden Rice, could provide greater than 50 percent of the recommended daily intake of vitamin A in these children, which could significantly improve health outcomes if adopted as a dietary regimen."

As one might expect, Greenpeace paid absolutely no attention to this categorical and unequivocal reaffirmation and defense of the actual research results. Rather, the organization simply passed over the issue in silence, as if the scientific facts didn't matter or didn't exist.

But this was by no means the end of the scandal. The *American Journal of Clinical Nutrition*, which had published Tang's paper describing the results, requested that she and the other authors voluntarily retract it, saying that otherwise the journal itself would unilaterally retract the paper for "ethical breaches" in the conduct of the study. Tang refused to retract the piece, which she said had been downloaded more than 32,000 times, and suggested instead that a few changes be made in the text. In July 2014 she also filed suit against the American Society for Nutrition, publisher of the journal, arguing that retraction was equivalent to defamation, and she sought a preliminary injunction against the proposed retraction, plus damages for defamation, breach of contract, and interference with business relations.

While Tang's lawsuit was awaiting a court hearing, Adrian Dubock, executive secretary of the Golden Rice Humanitarian Board, wrote a paper, "The Politics of Golden Rice" (2014), in which he charged the Tufts IRB with several infractions and ethical lapses of its own, including that the board "forbade [Dr. Tang] from discussing her research at a European science conference," "found her guilty of not following rules which were only introduced after the field research was completed," "advised her to withdraw her application for promotion during the investigation, with no explanation. And subsequently withheld about 50% of her pay for 2 months, claiming 'budget difficulties.'"

Finally, on July 16, 2015, a Massachusetts superior court ruled against Guangwen Tang and denied her motion for preliminary injunction, arguing that "the requested order would be an unconstitutional Prior restraint on speech as well as an unconstitutional order compelling speech." On July 29, the American Society for Nutrition retracted the paper, adding an explanatory note that said in part, "The authors are unable to provide sufficient evidence that the study had been reviewed and approved by a local ethics committee in China in a manner fully consistent with NIH guidelines" and that "they are unable to substantiate through documentary evidence that all parents or children involved in the study were provided with the full consent form for the study."

In response to the retraction, Greenpeace East Asia issued a victory declaration, saying in a press release dated August 11, 2015, "Twenty years after it was first conceptualized, GE 'Golden Rice' continues to be a failed experiment."

But neither Tang's experiment nor the rice itself had failed. As the three separate reviews launched by Tufts had unanimously concluded, Tang's research results had been reported accurately, and the rice itself had performed successfully and exactly as advertised.

"The study has shown that Golden Rice is a very effective source of vitamin A," Ingo Potrykus said of the incident. "That's what's most important."

And so Golden Rice remained untouched and unaffected by the one big scandal, which was about people and their actions, and not about the rice.

THE MISTAKE

And now, mirabile dictu, we have reached a point in the narrative for which practically nothing in the previous series of events has prepared us. This is, if not the climax of the story, then at least a miniclimax, a turning point, a high-water mark of the operation thus far. For this is the point at which Golden Rice finally emerges from the lab, from the greenhouse, from tiny confined field trials, into the full glare of day and into several different local indica rice varieties in three different countries. At long last, the rice is apparently inching its way into the hands of farmers and to its ultimate users, the poor and malnourished vitamin A–deficient populations of South and Southeast Asia.

There are some surprising successes to report . . . but also some failures that have to be acknowledged as spectacular. And, if this were politics rather than science, it would be the point at which prominent actors in the drama would utter the time-honored and timeless phrase, "Mistakes were made."

The story begins in March 2009, less than a year after the end of the 2008 confined field trial of GR 1 at the International Rice Research Institute in the Philippines, the planting that had been destroyed by a typhoon just a week before harvest. Not far from that field, the Golden Rice Humanitarian Board held a meeting at IRRI headquarters for what would turn out to be what its members themselves referred to afterward as a "watershed moment" in the development of Golden Rice.

It was now 10 years since the invention of the prototype, GR 0.0. Since then, Syngenta scientists, working also with Peter Beyer, had produced two new and improved versions of the rice, GR 1 and GR 2, both of which accumulated higher levels of beta carotene in the endosperm than the original. There were 23 versions of GR 2 alone, all expressed in the American japonica rice variety Kaybonnet, none in the indica varieties grown in Asia. The next step in the process of making the rice suitable for growth in Asian soils was to transfer the Golden Rice trait out of the japonica type and into preferred indicia rice varieties. That was done by conventional rice breeding techniques, by means of standard backcrossing, a process that had already begun at IRRI under the supervision of Parminder Virk.

But the Humanitarian Board members were facing a crucial decision, one that was forced upon them by the regulations that had kept the rice essentially a lab plant for almost its entire lifespan. The problem was that they had several different GR 1 and GR 2 transformation events in front of them, from which they had to choose just a single one to go forward as the *one lead event everywhere.*

And they had to do so, moreover, without exactly an overabundance of relevant information. But at the time they made their decision they were convinced that they had at least enough information to proceed. First of all there were datasets that had been produced by the scientific and breeding staff members of the Golden Rice program at IRRI. These datasets pertained to the three Cocodrie GR 1 transformation events and the six Kaybonnet GR 2 events that IRRI had received from Syngenta and had then transferred into different indica rice varieties. The information that the scientists derived from these breeding experiments covered a number of agronomic measurements that were valuable to rice breeders, along with data regarding carotenoid levels, as well as the degree of degradation of the carotenoids across time. But all these indicators came from plants that had been grown in greenhouses, and consequently they were subject to all the limitations pertaining to data from plants raised in such synthetic and controlled laboratory conditions.

Golden Rice had also been grown in India, but, according to Adrian Dubock, who took detailed notes on the 2009 "watershed" Humanitarian Board meeting, the data "was only available from the phytotron, due to the regulations governing GMO-crops, and the phenotypes were so adversely affected by the artificial environment that useful data could not be generated." S. R. Rao, the board member from India, "asked whether there was any molecular data available to support the decision making. No such data was forthcoming (although IRRI had received it in 2006, it appeared to have been forgotten)."

One gets the impression at this point that the members were in the position of being forced to make a momentous decision amid an actual paucity, or even a deficit, of adequate information. But there *was* some real data in front of them. For example, there was data on the bioconversion ratio of Golden Rice beta carotene to vitamin A. Just prior to the Humanitarian Board meeting, Guangwen Tang had provided the board members with the manuscript of her paper describing the first study of the bioavailability of Golden Rice in humans, the trial at Tufts involving five adult subjects from the Boston area. The results she reported there, taken together with the known beta carotene levels expressed in the kernels of GR 1 and GR 2, showed that the amount of beta carotene required to alleviate vitamin A deficiency would not be available from eating a reasonably-sized portion of GR 1. The lead transformation event, therefore, would have to be selected from the group of the six GR 2 events for which IRRI had generated greenhouse data.

After discussion, the board members decided that GR2G would be the lead event, with GR2R held in reserve as a backup if needed. The rationale for this choice was later explained by Adrian Dubock: "The reserve event GR2R performed better agronomically and from a ß-carotene accumulation perspective than the event GR2G selected as the lead. The reason, in these circumstances that GR2G was selected was because it has been used in the human bioconversion trials."

This was a somewhat curious rationale. Nevertheless, the fact that GR2G had been used in the bioconversion trials gave it a regulatory advantage insofar as it had already been consumed by humans, who had successfully converted its beta carotene into vitamin A at an experimentally established and impressive rate. In other words, GR2G was a tested, proven, and known entity, and furthermore it contained enough beta carotene so that a young child's vitamin A requirements would be met by consuming 100 grams (3.5 ounces) of the rice per day.

Using that as a basis for selection of the lead event, however, meant that regulatory compliance issues were once again driving scientific choices. This was a case of the tail wagging the dog, but such a policy had been forced upon the Humanitarian Board members by regulations that made real-world field trials so expensive and required amassing so much data in advance that doing all the necessary work for putting forward several transformation events simultaneously was simply out of the question financially. Thus the need for one lead selection event, and the associated need for a lead event that would be acceptable to regulators without further undue delay.

Within an hour of the Humanitarian Board reaching this decision, members of the separate Golden Rice Network met to consider the practical details of implementing it. The meeting was attended by representatives of the rice research institutes in Bangladesh, India, Indonesia, and Vietnam. Ingo Potrykus told the group that the lead event was GR2G, as well as the reasoning behind the choice. In turn, the rice scientists from the different countries presented their plans to breed that event into one or more specific mega-varieties that were locally favored and appropriate to their respective soils and climates.

Soon afterward, three countries had active breeding programs going. In the Philippines, PhilRice, the government-sponsored rice research institute was now working in conjunction with IRRI and breeding the GR2G event into a popular rice variety Rc-82 (also known as Peñaranda), a type that was responsible for about 13 percent of Philippine rice production. In Bangladesh, scientists at the Bangladesh Rice Research Institute (BRRI) in Gazipur, were introgressing the GR2G event into their single most important boro rice variety, Br-29. And, in India, a group of researchers at the Indian Agricultural Research Institute in New Delhi was modifying three different types with GR2G, including the highly popular rice type, Swarna. And so, after 10 years of research and many delays, the Golden Rice trait was at long last slowly being introduced into some of the world's major rice varieties, which was a necessary step in the process of getting it into the hands of rice farmers.

At which point, fate intervened.

In October 2009, seven months after the "watershed" meeting of the Humanitarian Board in March of that year, IRRI researchers learned that the sequencing of the inserted gene in their chosen lead event, GR2G, was incomplete. The mere lack of sequencing information, they thought, was functionally insignificant insofar as the plant's actual performance was concerned, but it would be a problem for the overall Golden Rice program because the incomplete information would invariably raise questions in the minds of regulators, for whom a data gap of any kind was a red flag, and so it would delay their review and ultimate approval of the product.

For a variety of technical reasons, the IRRI Golden Rice scientists now proposed to change the lead transformation event from GR2G to the backup event, GR2R (which Dubock had acknowledged "performed better agronomically and from a ß-carotene accumulation perspective") and to put forward yet a *third* event, GR2E, to be held in reserve as a new backup. According to Dubock, since the Humanitarian Board members "had not reviewed, nor did most have the training, to 'review all sequence data' in any meaningful way," the members were forced to rely on the judgment of the IRRI scientists. In December 2009, therefore, the Humanitarian Board unanimously accepted the IRRI recommendations. GR2R would be the new lead transformation event, and GR2E the new backup.

The long and bumpy road of Golden Rice research and development had been littered with endless obstacles: failed experiments, false starts, detours, unpleasant surprises, puzzling results, unanswered questions, unsolved problems, and multiple uncertainties and unknowns. However, all of that and more were fairly standard problems faced by those doing cutting-edge, original scientific research, which the process of inventing Golden Rice certainly was.

Still, the choice of GR2R as the new lead event would prove to be the One Big Mistake.

■ Once the lead event had been changed to GR2R, the three countries (Philippines, Bangladesh, and India) that had started breeding the old event, GR2G, now switched their breeding programs over to the new one.

In the Philippines, GR2R seeds went to PhilRice. Under PhilRice project leader Antonio A. Alfonso, breeders introgressed the GR2R trait into the old standby, IR64 rice, producing a variety that was technically designated as IR64XGR2-R. It took the scientists three years to fully add the trait into the IR64 cultivar, which was exactly the amount of time that Gurdev Khush had long ago said it would take to produce an Asian-adapted Golden Rice variety from the original prototype.

In the end, the breeders created nine slightly different versions of the IR64 variety and tested them first in screenhouses. Then, in 2012, after having secured approval from the Philippine Department of Agriculture, the scientists conducted a series of multilocation field trials at five sites around the country, to evaluate the agricultural performance of the rice plants in various real-world sites, soils, and weather conditions.

Each test site was to be no more than 1,000 square meters in size (about a quarter of an acre), surrounded by a tall fence, and patrolled day and night by security guards. As was required by regulations, the sites of all field trials were announced online in advance, making them easy for anyone to find.

During the first round of multilocation field trials, the yield of the new Golden Rice variety was slightly less than that from the ordinary, non-GMO IR64 plants grown alongside as controls. Total carotenoid content ranged between 4.51 and 5.55 micrograms per gram—acceptable but hardly impressive numbers.

At the end of the trials, also as required by biosafety guidelines, all excess grains were heat-killed and then buried inside the fenced area, after the manner of hazardous materials or radioactive waste disposal. Even after those precautions were taken, the area was monitored for a month to guard against the possibility of any heat-killed Golden Rice material resurrecting itself and rising from the ashes.

The Philippine scientists conducted a second round of multilocation field trials the following season, in 2013. On August 8, just before the plants were ready to be harvested, a large group of vandals broke through the protective fence and destroyed the crop of Golden Rice plants at the Pili field trial location, a site in the province of Camarines Sur.

The attack was reported in the world's press—and often misreported. *New Scientist*, for example, attributed the destruction to a group of 400 "militant Filipino farmers." But Filipino farmers, who harbored an almost mystical respect for rice plants, would not have trampled a growing crop. Raul Boncodin, a Golden Rice project manager at IRRI, who witnessed the attack, said, "Maybe two or three of them were farmers, but the rest of them were not. . . . They were city boys, city girls. Two of them were even sporting dyed hair. Would you consider a farmer having dyed hair?"

In an official statement, the Philippine Department of Agriculture said the attack had been staged by a coalition of local anti-GMO activists. There was no direct evidence that Greenpeace was involved in the vandalism, and the organization took no credit for it, although it continued to criticize Golden Rice. The destruction of these crops was condemned by the global scientific community, and one petition of support for further Golden Rice research and field trials gathered a total of 6,772 signatures.

Quite apart from the vandalism, the results of the multilocation field trials proved to be a disappointment for Golden Rice. In May 2014 IRRI reported on its website that "preliminary results were mixed. While the target level of beta-carotene in the grain was attained, average yield was unfortunately lower than that from comparable local varieties already preferred by farmers. . . . Based on these results, a decision has been reached to move forward from work solely focused on GR2-R to also include other versions of Golden Rice, such as GR2-E."

Critics of Golden Rice blamed the poor performance on an inherent deficiency in molecular technology that inevitably produced defects in the rice. Defenders blamed the government regulations that made the scientists choose a lead event in the absence of sufficient knowledge to make the proper choice. Both sides ultimately agreed that choosing GR2R as the lead event had been a mistake.

GR2E was the backup version of GR2R. It was now the *third* lead event in sequence, and it was this one that would in fact ultimately succeed. But implementing that event would take another three years of backcrossing—or perhaps less by using the technique of marker-assisted back-crossing—plus field trials and evaluation of results.

■ Initially, Golden Rice researchers in Bangladesh also worked with the second lead event, GR2R. The breeding was conducted at the Bangladesh Rice Research Institute, at Gazipur, under the direction of principal investigator Partha S. Biswas. In this case, the scientists introgressed the GR2R trait into the local variety BRRI Dhan 29, giving a rice type formally referred to by the longish name, GR-2R BRRI Dhan 29. By 2013, the new variety was ready for a confined field trial at a single location in Gazipur.

The official application for the field trial shows in detail the extent of the precautions taken for the open-air testing of this "regulated article." GR-2R BRRI Dhan 29 rice would be planted in only three rows, in a plot that was 900 square meters in size and surrounded by fencing. No other rice variety would be planted within 200 feet of the site, but five rows of corn plants would be grown around the area "to avoid pollen flow."

As a further precaution, "a double drainage system improvised with PVC pipe capped with spider net will be used to avoid outflow of seeds and vegetative parts with water from drainage channel. The drained water will be allowed [to] deposit in a pit inside experimental site."

Security measures at the location were comparable to what might be found at a secret government nuclear weapons center. The test site would be continuously attended by security guards and would be covered with netting to keep stray birds away from the rice plants. Access to the area would be limited to authorized persons who would be carefully logged in and out. At the end of the test, all vegetative parts, excess seeds, stubble, and other plant debris would be burned and then plowed under, and the site would be monitored for 45 days. Any plants that might afterward emerge from the fallow area would be shoveled underground or pulled out by hand and destroyed.

Partha Biswas and his breeders had developed 22 lines of GR-2R BRRI Dhan 29, and from them they selected 7 lines for field testing. At the end of the trials in 2013–2014, the tested lines showed up to 12 percent yield advantage over regular, non-GMO BRRI Dhan 29. Total carotenoid levels, as analyzed after three months of storage varied from 8.1 to 15.5 µg/g.

That was the good news. The bad news, such as it was, had long since been communicated to the growers by Peter Beyer, which was that in view of its poor performance at IRRI in 2013, GR2R had been replaced by GR2E as the lead event. Once the R event had been exposed to real-world conditions, it supplied the scientists with actual empirical data, which is exactly what they had needed to make the lead event selection on a sufficiently informed basis in the first place. Now that they had the additional information, at least pertaining to the R event, they could switch to the backup E event and hope for the best.

The switchover was no problem for Biswas, as his crew simply started the entire process all over again with the E event, producing yet another new rice line, GR-2E BRRI Dhan 29. By 2015, eight separate lines were ready for testing in Bangladesh, and, perhaps not surprisingly, they produced even better results than the previous version. The new Golden Rice plants, their panicles (seed clusters), and grains were morphologically very similar to their non-transgenic counterparts, with comparable plant height, yield, grain weight, shape, and size. Total carotenoid content ranged from 3.5 to 26.3 micrograms per gram, the latter of which was a very high level of expression. This, finally, was a rare and completely unexpected outcome for Golden Rice: an apparently resounding and unqualified success.

Emboldened by this triumph, in 2016 Biswas and his breeders repeated the field trial and got similar positive results. "Two months after harvest, we've found an average of over 10 µg/g beta carotene in GR2E BRRI Dhan 29," Biswas reported. "The amount is good enough to meet fifty percent of vitamin A needs of people consuming rice in their daily diet."

Biswas was planning more extensive, multilocation field trials at five sites across the country for 2017 and 2018. His hope was that government approval for commercial release to farmers would follow soon afterward. If that indeed happens by 2019, it will have been 20 years since the announcement of the Golden Rice prototype in 1999.

■ Matters were not so promising in India, however. After IRRI recommended and the Golden Rice Humanitarian Board accepted the selection of GR2R as the lead event, a group of researchers at IARI in New Delhi took delivery of GR2R seeds as well as seeds of the five other transformation events that Syngenta had released to the Humanitarian Board for use in the public-sector breeding programs.

Under the direction of Ashok Kumar Singh, head of the IARI Division of Genetics, members of a research group began the process of backcrossing the beta carotene trait into a local mega-variety, a strain called Swarna, India's most widely grown rice. The researchers used marker-assisted backcross breeding, which is faster than conventional breeding, and so it did not take them the usual three years to introgress the beta carotene trait into Swarna.

They started breeding the R event in 2009 and initially got some good results, "with very high level of total carotenoid expression, i.e., 25 µg/g," according to Singh. However, in 2011 the group began a second round of breeding, and the results in this case were disappointing. In fact, they were rather alarming. While the transformed plants produced high levels of beta carotene, just as before, the plants themselves were characterized by "several aberrations in their phenotype" as compared to normal Swarna specimens. Specifically, the plants were so short in stature that they could be called "dwarf" and produced "pale green leaves and drastically reduced panicle size, grain number and yield as compared to the recurrent parent, Swarna." The roots of the plants did not develop normally, either: they were longer than usual and failed to grow vertically along the typical downward path reflecting the force of gravity, instead branching out horizontally, which was anomalous. By any ordinary and reasonable criterion, this plant line was a shambles.

Faced with this botanical disaster, the scientists had two alternatives: One, they could abandon the R event and immediately proceed to the E event, as Partha Biswas had done in Bangladesh. Two, they could forget about the E event, at least temporarily, and instead take up the perhaps more interesting scientific challenge of investigating and explaining the molecular basis of the abnormalities they had observed in the performance of the R event. They chose the latter.

"In September 2011 Ms. Haritha Bollinedi joined as my Ph.D. student and started working on it," Singh reported. "And by the middle of 2012 we knew that *Aux1* disruption by transgene insertion was the reason for this abnormality."

Aux1 is a gene that codes for a protein that regulates the growth, development, and gravitropic curvature of the plant's roots. The disruption was a consequence of the genetic engineering method that the Syngenta scientists had used to produce SGR2, *Agrobacterium*-mediated transformation, which leads to random insertion of transgenes in the plant genome. In this case, the inserted gene disrupted the function of certain plant hormones and produced other downstream effects.

The scientists theorized that the further abnormalities had resulted from other components of the transgene used in GR2R which caused alterations to different parts of the engineered biosynthetic pathway. "Any change in one pathway," they said, "would influence the others as well."

At the end of their investigation, the scientists produced a detailed report of their findings in a paper written mainly by Bollenedi and published in 2017 in the online journal *PLoS ONE*:

> The ideal transgenic plant for most research and breeding purposes would contain a single intact copy of the desired transgene inserted into the host genome, such that it does not alter the functionality of the host plant DNA. *Agrobacterium*-mediated transformation leads to random insertion of the transgenes in the genome. Therefore, it is very important to determine the genomic location of the transgene insertion. . . .
>
> Based on the morphological, biochemical and molecular characterization of backcross derived transgenic Golden Rice lines in the genetic background of Swarna, we discovered that the

insertion of transgene for pro-vitamin A trait in the donor GR2-R1 event in Kaybonnet, had disrupted a native *OsAux1* gene, which resulted in phenotypic abnormality and poor agronomic performance of the backcross derived Golden Swarna lines, making them unfit for commercial cultivation inspite of having high provitamin A content. The study conclusively demonstrates the importance of event characterization and event selection before adopting the transgenic germplasm into introgression breeding.

Potrykus, Beyer, and company could not agree more with that assessment. If they and the rest of the Golden Rice Humanitarian Board had had fuller information about the transformation events to begin with, it is likely that they never would have selected GR2R as the lead event.

■ The collapse of GR2R Swarna was a worst-case example of the unintended negative consequences so often predicted by GMO critics. Indeed, the Swarna result was an anti-GMO activist's dream come true, and opponents of genetically engineered foods, in particular opponents of Golden Rice, quickly leaped to exploit what might be called a Golden Opportunity to trash the product. They portrayed what had happened in the most damaging possible terms, made claims that far exceeded the actual evidence, and forecast the occurrence of the wildest and direst consequences had the rice been distributed to farmers and grown. In the process they essentially pronounced a death sentence upon Golden Rice, R.I.P.

For example, in 2017 Jonathan Latham, a GMO critic who was also the executive director of the Bioscience Resource Project, said, "What the Indian researchers show is that the Golden Rice transgenes given to them by Syngenta caused a metabolic meltdown. The classic criticisms of genetic engineering as a plant breeding tool have always been, first, that introduced DNA will disrupt the native gene sequences and, second, that unpredictable disruption of normal metabolism may result from introducing new functions. Golden Rice exemplifies these flaws to perfection."

Latham really should have known better than to disparage "genetic engineering as a plant breeding tool" on the basis of the failure of a single Golden Rice event that had been introgressed into a single selected rice type, Swarna. "The anti-GM lobby has misinterpreted our findings as a failure of GR technology," Ashok Singh has said. "One must understand that the problem with the R event is event-specific and nothing to do with GR technology per se."*

In other words, there was no basis in the Swarna R-event results to draw any wider conclusions about any other Golden Rice event or rice type or to dismiss Golden Rice, or genetic engineering in general, in Latham's sweeping and wholesale manner. The GR2R event in Bangladesh, as bred into a different rice type, BRRI Dhan 29, suffered no "metabolic meltdown" or indeed any other adverse consequence. It grew phenotypically normal and healthy plants that produced beta carotene precisely as they had been designed and genetically engineered to do. Those plants exemplified the virtues, not the "flaws," of genetic engineering.

There was also no basis in the failure of the event-specific Swarna R event to condemn genetic engineering methods as a whole. The fact is that the kinds of genomic disruptions that Latham mentions as if they were exclusive to genetic engineering also occur in traditional plant breeding. Introduced DNA disrupts gene sequences in *all* forms of plant breeding. Indeed, such disruptions occur in nature during the course of the random and natural gene mutations that drive evolution.

Creating a new plant variety or introducing a new desired trait in an existing variety requires making a mutation in the parent plant, and to make a mutation is, as Latham says, to "disrupt the native gene sequences." But such genomic disruptions, far from being unique to molecular breeding methods, are simply the means by which new variants of any kind of plant or animal are brought into existence. If and when those genomic disruptions cause unintended negative effects in traditional plant breeding, the plant line in question is normally discarded by the breeder or farmer as part of the ordinary course of events in any plant breeding program.

* Ashok Singh, email to Ed Regis, December 29, 2017.

Moreover, a body of research evidence shows that genetic engineering methods in fact cause *less* disruption to the host genome than do traditional plant breeding methods. A 2006 study by a group of British scientists compared the gene expression profiles of transgenic versus conventionally bred lines of wheat. By the use of microarray analysis the authors determined that conventional breeding methods introduced *greater* changes in gene expression than did genetic engineering techniques.

"Differences in observed gene expression in the endosperm between conventionally bred material were much larger in comparison to differences between transgenic and untransformed lines," the authors wrote. "These results suggest that the presence of the transgenes did not significantly alter gene expression and that, at this level of investigation, transgenic plants could be considered substantially equivalent to untransformed parental lines."

A later series of experiments produced a similar conclusion, specifically regarding rice. In a 2008 study published in the *Proceedings of the National Academy of Sciences*, a team of European researchers compared the effects on four rice lines modified by mutation (gamma radiation) breeding to the effects on the plants produced by genetic engineering. They found that both breeding strategies cause stresses to the plants and furthermore that the stressing effects are passed down to succeeding plant generations, although there are fewer such effects in each new generation. The surprise was that genetic engineering methods caused *less* stress to the plants than mutation breeding did.

> We found that improvement of a plant variety through the acquisition of a new desired trait, using either mutagenesis or transgenesis, may cause stress and thus lead to an altered expression of untargeted genes. In all of the cases studied, the observed alteration was more extensive in mutagenized than in transgenic plants. We propose that the safety assessment of improved plant varieties should be carried out on a case-by-case basis and not simply restricted to foods obtained through genetic engineering.

Further, introduced genomic disruptions are in and of themselves neither good nor bad; rather, it all depends on whether the consequences of the disruptions are desirable or undesirable. In the case of GR2R Swarna, the consequences were bad; but in the case of both GR2R and GR2E BRRI Dhan 29 as field-tested in Bangladesh, they were good: the plants were structurally sound and the kernels produced beta carotene at levels that could help reduce vitamin A deficiency.

A further criticism leveled at Golden Rice by GMO opponents was that the deformed Swarna R plants presented a danger to rice farmers everywhere because they could wreak havoc on normal, untransformed rice plants.

Christopher Then, a GMO critic formerly with Greenpeace, argued, "Once released, the transgenic plants could spread their gene constructs into populations of weedy rice as well as other cultivated varieties. Instead of helping people to combat malnutrition, these plants, if grown on the fields, might endanger their whole rice harvest."

To the contrary, plants that are as metabolically challenged as the Swarna R events would be at such a selective disadvantage with respect to untransformed rice populations that competitive pressures would almost certainly drive the metabolically melted-down R variety to a swift, utter, and deserved extinction.

But all these points were lost on GMO critics motivated by the single-minded goal of seizing upon the failure of the R event in Swarna and using it to dismiss Golden Rice in general and to prophesy its doom. And so the anti-GMO literature and blogosphere witnessed a succession of articles and headlines proclaiming: "Golden Rice Flops," "Goodbye to Golden Rice," "The End for Golden Rice," and so on. But those death sentences were premature.

■ Because now, just as the scientists at the Indian Agricultural Research Institute were reporting on the failure of GR2R Swarna, and while a Greek chorus of GMO critics celebrated the imminent demise and disappearance of Golden Rice from the scene, the plant itself was experiencing what amounted to a miraculous return to life and health in the Philippines.

In 2014, after IRRI's decision to abandon its low-yield GR2R IR64, PhilRice and IRRI jointly undertook a program of breeding GR2E, the third lead event, into three distinct rice varieties. These were the workhorse IR64, the Bangladeshi BRRI Dhan 29, and the Philippine PSB Rc82. Then, in the rainy (summer) season of 2015 and in the dry (winter) season of 2016, the scientists conducted field trials of those three transformed lines at four locations in the Philippines. The researchers took regular weather observations and recorded monthly temperature, humidity, wind speed, and precipitation data for each site across the entire duration of the tests. They kept detailed accounts of the maintenance products (fertilizer, pesticides) used, as well as a record of dates of application and in precisely what amounts. And they monitored growth stages of the plants, noting the times of panicle initiation, flowering, and dates of harvesting, among other developmental stages.

After harvest, the scientists undertook the usual compositional analysis of the rice grains produced in the trials to determine their total carotenoid content. And from this measurement emerged two important facts. First, each of the three lines produced significant levels of carotenoid, in concentrations sufficient to be helpful in alleviating vitamin A deficiency. Second, of the three varieties, one was a clear winner and the star of the show. This was BRRI Dhan 29 GR2E rice, which exhibited mean total carotenoid values lying between 24.5 and 29.33 micrograms per gram.

All three lines exhibited another desired feature as well: repeated plantings produced four successive generations of each rice type, and these offspring plants showed that the beta carotene trait was inherited from generation to generation, at consistent and stable levels of expression. In sum, these were healthy, robust plants that bred true. Clearly, a milestone had been reached, and surpassed, in the development of Golden Rice.

At this point, PhilRice and IRRI were prepared to take the next major step, which was to submit to the Philippine Department of Agriculture an application for a biosafety permit seeking official government approval for Golden Rice and to allow the use of GR2E rice as food and

feed in the Philippines. Almost 20 years of hard work was about to cul-
minate in the first such formal application for approval and release.

The application was submitted in 2017, and consisted of an imposing
set of eight separate documents. One of them, *Provitamin A Biofortified
Rice Event GR2E*, was a 138-page report that functioned as an overview
and summary of the entire project. The document included detailed information on the use history of the host and donor organisms; the genetic modification and transformation method used; a molecular genetic
characterization of the rice; plus additional sections on the safety of the
proteins used in the transformation construct, as well as a compositional
breakdown of the various minerals, amino acids, fatty acids, and vitamins in the rice endosperm. The compositional breakdown demonstrated that there existed no foreign or unintended elements in the endosperm, and that in fact the only new molecules in the grains were those
that had been intentionally introduced by the genetic transformation.

More impressive still was the *Compilation of Study Reports: Studies
Submitted in Support of the Food Safety Assessment of Provitamin A Biofortified GR2E Rice*. This was an 872-page data-rich compendium of information about virtually all known physical and nutritional characteristics of the rice and its molecular components, plus a discussion of the
various allergenicity and toxicity screenings to which GR2E rice had
been subjected.

The PhilRice/IRRI scientists stated their final conclusions about
Golden Rice in three succinct paragraphs:

> The purpose of this evaluation of GR2E rice was to determine
> whether the use of GR2E rice in food or feed could raise any new
> safety concerns relative to conventional rice, and was not intended
> to address questions related to the efficacy of GR2E rice in helping
> combat VAD in at-risk populations.
>
> The assessment of GR2E rice has included a complete descrip-
> tion of the genetic modification (e.g., gene sources, characterization
> of inserted DNA and site of integration within the host genome,
> stability, and inheritance), the safety of the newly expressed pro-

teins (e.g., history of use in food, function, potential toxicity, potential allergenicity, and patterns and levels of expression), and a comprehensive nutrient compositional assessment to identify whether there were any unintended, unexpected, effects of the genetic modification.

Collectively, the data presented in this submission have not identified potential health and safety concerns, and support the conclusion that food and/or livestock animal feed derived from provitamin A enriched GR2E rice is as safe and nutritious as food or feed derived from conventional rice varieties.

These two documents, which together totaled more than 1,000 pages, constituted a graphic, voluminous, and weighty proof of the claim, often made by Golden Rice proponents, that their provitamin A biofortified rice was one of the most exhaustively characterized and well-studied food products in the history of nutrition.

Every indication now was that, against all odds, Golden Rice was finally moving toward deregulation, farmers' fields, and to children and pregnant women suffering from VAD. This was a product that had taken six years of lab work to get to the proof of concept. The rice had to overcome numerous scientific challenges and extremely burdensome regulatory obstacles. It had to withstand years of constant abuse, opposition, factual distortions, disinformation, and ridicule by anti-GMO individuals and groups. It had to survive the destruction of field test specimens by cyclones, hurricanes, and paid vandals. It had to survive its one major scandal and one major mistake.

That in the face of all this it was now on the verge of official government approval was almost unbelievable.

Yet in another sense it was quite understandable. After all, Golden Rice was just rice. Just plain and ordinary rice. Only healthier.

THE "CRIME AGAINST HUMANITY"

Patrick Moore, one of Greenpeace's founding members and, much later, director and, still later, dropout, critic, and opponent of the organization, first met Ingo Potrykus at a conference in Helsinki, Finland, not long after the invention of the Golden Rice prototype. At that point, Potrykus was already involved in the thankless process of responding to Greenpeace attacks on him, his research, Golden Rice, and GMOs in general. The time was already past when Benedikt Haerlin had visited Potrykus at his home in Switzerland and calmly discussed whether to make the provitamin A rice an exception to Greenpeace's otherwise absolute and rigid opposition to any and all genetically engineered foods. Haerlin had initially acknowledged the moral difference between GMOs that were merely agriculturally superior—in being inherently pesticide or herbicide resistant, for example—and a GMO that was so nutritionally beneficial that it actually had the potential to save people's lives and sight. But apparently that distinction made no difference to Haerlin, because in the end both he and Greenpeace itself soon reverted to the view that Golden Rice had to be opposed, even stopped, no matter what its possible health benefits might be.

On February 9, 2001, Greenpeace issued its "'Golden Rice' is Fool's Gold" press release, and a few days later, on February 15, Potrykus published a written response, titled "Golden Rice and the Greenpeace Dilemma." He pointed out how the Greenpeace press release had misquoted him, how it had misrepresented the likely bioavailability of the beta carotene in the rice, and how Greenpeace spokespeople had failed, despite numerous requests, to offer any plausible rationale for their repeated claims that field trials of the rice would be harmful to the environment by "contaminating" other crops or successfully competing against and displacing conventional rice plants.

"As the [beta carotene] pathway is already in rice (and in every green plant)," Potrykus wrote, "and the difference is only in its activity in the endosperm, it is very hard to construct any selective advantage for Golden Rice in any environment, and, therefore, any environmental hazard."

However, and to be fair, the fact that Golden Rice had no selective advantage that would make it an environmental threat doesn't by itself mean that the rice couldn't contaminate other rice varieties. Golden Rice proponents have often claimed that, because rice is self-pollinating, there is only a low risk of cross-pollination or of Golden Rice contaminating other varieties. But the fact that it is unlikely doesn't mean it can't happen. It can, and in fact it has—although not (yet) with Golden Rice.

In the late 1990s, the biotech firm Aventis—the same firm that had created StarLink corn for use as animal feed, which became inadvertently mixed with human foods and had to be recalled—also invented a new rice variety, Liberty Link. The company had altered the genes of a strain of long-grain rice to resist the weed killer Liberty, which itself was manufactured by Aventis. There were two lines of Liberty Link rice, LLRICE601 and LLRICE604, neither of which had been "deregulated," meaning approved for use as foods in the United States.

In 2001, Aventis sold its agrochemicals unit to Bayer CropScience AG, in Germany, which then took over the Liberty Link rice lines. In 2006 Bayer reported that the regulated genetic material in LLRICE601 had been detected in the American long-grain rice variety Cheniere. And in 2007 the regulated genes in LLRICE604 had been detected in another American long-grain variety, Clearfield 131.

The USDA immediately launched a massive investigation in which it tested 396 samples of 57 rice varieties from 11 US states and Puerto Rico. After more than 8,500 staff hours of work, the Department of Agriculture issued a report on the incident in which three conclusions were noteworthy.

One, "investigators had hoped to identify how each GE rice line entered the commercial rice supply, but the exact mechanism for introduction could not be determined in either instance." The report stated that such contamination "can result from natural processes such as the movement of seeds or pollen or human-mediated processes associated with field testing, plant breeding, or seed production." But the investigators couldn't say what the method of transmission had been in this case, and it remains unknown to this day.

Two, the mixing of genetic information occurred "at extremely low levels—of genes and gene products from unintended plant sources. This can occur with both conventionally bred plants as well as biotechnology-derived plants."

This is an important point. It means that there is nothing special about genetically engineered plants that makes them more likely than conventionally bred plant types to contaminate other varieties of the same plant. The scenarios routinely presented by Golden Rice critics invariably portrayed Golden Rice as taking over other rice types and blotting them out of existence, whereas in fact it would be just as likely for those other types to infiltrate and "contaminate" Golden Rice. But that possibility is one that the critics never mention.

Three, "the GE rice poses no identifiable concerns related to agriculture or the environment." Which means that Golden Rice itself would not pose any such risks either.

Nevertheless, despite the improbability of Golden Rice contaminating conventional rice, Greenpeace went so far as to state that, because of the environmental threat it posed, the organization would take direct action to stop Golden Rice from being grown, to the point of ripping the plants out of the ground if necessary. In response Potrykus issued a warning of his own: "If you plan to destroy test fields to prevent responsible testing and development of Golden Rice for humanitarian purposes, you will be accused of contributing to a crime against humanity. Your actions will be carefully registered and you will, hopefully, have the opportunity to defend your illegal and immoral actions in front of an international court."

Patrick Moore liked that idea. "I wholeheartedly agreed with him and seconded the motion," he recalled later. "Potrykus is correct; they should stand trial for crimes against humanity."

The notion that interfering with the testing, production, or distribution of Golden Rice amounts to a crime against humanity has a long history. The view is at least superficially plausible. After all, if a million children per year suffer blindness and preventable death due to vitamin A deficiency, and if Golden Rice could help alleviate that deficiency and prevent at least some of those illnesses and deaths, then any efforts to halt, delay, or derail Golden Rice must be criminal.

Such a claim has been made again and again over the years by various GMO scientists and supporters, in tandem with the constant cycle of delays in the approval and release of Golden Rice. All of it culminated in June 2016, when more than 100 Nobel laureates, including James Watson, codiscoverer of the structure of DNA, signed the "Letter Supporting Precision Agriculture (GMOs)" and posted it on the website supportprecisionagriculture.org.

The letter was addressed "to the Leaders of Greenpeace, the United Nations and Governments around the word" and said, in part, "We urge Greenpeace and its supporters to re-examine the experience of farmers and consumers worldwide with crops and foods improved through biotechnology, recognize the findings of authoritative scientific bodies and regulatory agencies, and abandon their campaign against 'GMOs' in general and Golden Rice in particular." The letter continued with two calls to action and ended with a question:

WE CALL UPON GREENPEACE to cease and desist in its campaign against Golden Rice specifically, and crops and foods improved through biotechnology in general;

WE CALL UPON GOVERNMENTS OF THE WORLD to reject Greenpeace's campaign against Golden Rice specifically, and crops and foods improved through biotechnology in general, and to do everything in their power to oppose Greenpeace's actions and accelerate the access of farmers to all the tools of modern biology, especially seeds improved through biotechnology. Opposition based on emotion and dogma must be stopped.

How many poor people in the world must die before we consider this a **"crime against humanity"**?

The letter was major news, reported around the world under the appropriate headlines: "Stop Bashing G.M.O. Foods, More Than 100 Nobel Laureates Say" (*New York Times*), and "107 Nobel laureates sign letter blasting Greenpeace over GMOs" (*Washington Post*), and so on. But it all became yesterday's news soon enough, and of course the letter had no effect whatsoever upon Greenpeace other than to prompt it to issue a press release roundly denying all charges.

There was one person, however, who actually took action and who put together a case that Greenpeace was indeed guilty of crimes against humanity. And that was Patrick Moore.

■ Patrick Moore was so incensed by what he perceived to be Greenpeace's interference with the development of Golden Rice that he decided to exert some Greenpeace-style pressure against the very organization that he himself had played such a formative role in founding. In 2013, he, together with his brother Michael Moore, started a nonprofit activist group of their own, the Allow Golden Rice Now! Society, registered in British Columbia. Its purpose, according to its web page, was "to end the active blocking of Golden Rice by Greenpeace and other organizations who claim that it is either of no value or that it is a detriment to human health and the environment." The society aimed to do this in the same way Greenpeace itself had operated: by means of protests and demonstrations, media communications, and coalition building. The group would be funded by small donations raised locally and through the internet and by speaking fees.

It was in early August 2013 that vandals in the Philippines had destroyed the GR2R Golden Rice experimental plants at IRRI. By October of that year, Moore's Allow Society was already in action, staging a demonstration at a Vancouver wharf where a Greenpeace vessel, the *Rainbow*

Warrior, was on public display. Moore, his brother, and some of their children arrived at the dock and unfurled and held aloft a yellow banner bearing the words: "Greenpeace's Crime Against Humanity: 8 Million Children Dead." Speaking into a megaphone, Moore told the audience that had collected around him the bare-bones story of Golden Rice and how Greenpeace had consistently opposed it. On the ship's deck, meanwhile, Yossi Cadan, the campaigns director of Greenpeace Canada, watched the proceedings with some amusement, it being a new experience for him to be the object of a protest rather than a participant.

The next year, Moore made a European tour and held press conferences, gave lectures, and mounted protests at Greenpeace offices in Hamburg, Amsterdam, Brussels, Rome, and London. In 2015, he campaigned in Asia, touring the Philippines, Bangladesh, and India, lecturing, holding press conferences, and protesting at local Greenpeace offices. Outside the headquarters of Greenpeace Manila, Moore and a few others held up banners and chanted, "Greenpeace think twice, allow Golden Rice."

The turnouts for these events tended to be modest, however, and the various Allow activities seemed to have accomplished little other than a measure of publicity in the form of increased press and television coverage of the history, opposition to, and the forever-in-limbo status of Golden Rice. And, as a strictly practical matter, it was unclear how Greenpeace or anyone else opposing it could now "allow" Golden Rice since, at least during the years 2013 through 2015, there was no version of it in existence that had been successful enough in field trials to be considered for approval or release. There was no actual product either to allow or to prohibit.

As if in compliance with the action-and-reaction law of Newtonian physics, at about the same time that Moore's Allow organization became active there developed a countervailing group, the STOP Golden Rice Alliance (later the Stop Golden Rice! Network), based in the Philippines. The group described itself as "an alliance of farmers' groups, civil society organizations, sectoral groups and individuals calling to stop all forms of genetically modified crops, foods, and other tools of corporate control in food and agriculture."

In March 2015, just as Patrick Moore was launching his Asian tour, the STOP Alliance issued a statement opposing it: "The STOP Golden Rice Alliance strongly denounces the Golden Rice Campaign Tour planned for the Philippines, Bangladesh, and India initiated by the Allow Golden Rice Now! this March. Golden Rice is a covert attempt to win wider approval for genetically modified food and will not solve problems of malnutrition." This denunciation had no effect, either, and Moore completed his Asian tour as planned.

But over and above his publicity campaigns, it was a further part of Moore's ambition to formally charge Greenpeace with a crime against humanity. Laying out the basis of his accusation on the Allow Society's website, Moore started by citing and quoting from the Rome Statute of the International Criminal Court. The statute was a United Nations treaty that established the International Criminal Court at The Hague. The court would investigate and prosecute four core crimes: genocide, crimes against humanity, war crimes, and the crime of aggression.

The treaty had been adopted at a diplomatic conference in Rome in 1998 and was then opened for signature by any and all states. More than 100 of the world's nations signed the document, including Canada and the United States, and in July 2000 Canada also ratified the agreement. The United States did not, however, and in May 2002 notified the UN secretary-general of "its intention not to become a party to the treaty. Accordingly, the United States has no legal obligations arising from its signature on December 31, 2000." The treaty nevertheless entered into force on July 1, 2002.

Moore and his Allow Golden Rice Now! Society were in Canada, which *was* party to the treaty, and on the society's website Moore quoted two paragraphs from Article 7 of the Rome Statue, which covered crimes against humanity:

> For the purpose of this Statute, "crime against humanity" means any of the following acts when committed as part of a widespread or systematic attack directed against any civilian population, with knowledge of the attack: . . .

(k) Other inhumane acts of a similar character intentionally causing great suffering, or serious injury to body or to mental or physical health.

Moore then went on to claim that for 14 years Greenpeace had acted to prevent Golden Rice from being produced and reaching those who needed it. He stated further that "Greenpeace's actions are 'intentional' " and that its campaign is "a widespread or systematic attack directed against any civilian population, with knowledge of the attack." And he concluded by saying, "We are calling on the countries that are Party to the Rome Protocol to investigate this situation with an eye to taking Greenpeace before the International Criminal court to answer for their crime against humanity."

His call to action had no takers, and no state party to the Rome Statute has brought any such charge to the International Criminal Court at The Hague. Nor is any nation likely to. Indeed, Moore's case suffered from a fatal flaw: it ignored the critical distinction between Greenpeace's actions being *intentional*, in the sense of being done on purpose or consciously motivated (which they were), and done with the *intention of causing great suffering* (which arguably they were not). Further, even if it were true that Greenpeace's actions aimed at stopping GMOs had the *effect* of causing suffering, it would not follow that causing suffering was the *intent* of those actions. The suffering the actions caused, if any, would be an unintended, although foreseeable, consequence of their opposition to GMOs—more like collateral damage in the military sense, a case of inadvertently inflicting casualties on innocent civilians.

In making his argument, Moore stated, "There is no doubt that Greenpeace and its allies are largely if not entirely responsible for the opposition to Golden Rice that has effectively blocked its cultivation and delivery." But while Greenpeace and its allies might well constitute Golden Rice's main, most persistent, and most vocal foe, whether its opposition had "effectively blocked" any aspect of the rice's development was a highly dubious proposition. Greenpeace itself denied that charge: on June 30, 2016, in reply to the Nobel laureate "crime against humanity" letter, a Greenpeace Manila press release stated, "Accusations that any-one is blocking genetically engineered 'Golden' rice are false. 'Golden' rice has failed as a solution and isn't currently available for sale, even after more than 20 years of research."

But if Greenpeace *had* indeed "blocked" Golden Rice, how had it done so? It had ridiculed Golden Rice and called it names ("fool's gold," "all glitter, no gold," "more hype than substance," "propaganda for the genetic engineering industry," and a "golden illusion"). It had exaggerated the amount that was needed to meet the daily recommended requirement of vitamin A. It had branded Golden Rice a threat to the environment and to human health, without providing evidence for either claim. It said Golden Rice would not solve the "real problem" of vitamin A deficiency, which was poverty, and that it had the potential for causing "unintended effects."

But these and other such claims, whether true or false, did not by themselves block, impede, or retard Golden Rice development, nor did they in fact have the power to do so. None of these claims stopped Beyer and Potrykus from inventing Golden Rice in the first place, nor did they prevent Tran Thi Cuc Hoa, Peter Beyer, and others from developing GR 0.5, nor Beyer, Syngenta, and others from developing GR 1, nor Syn-genta from improving the rice further and creating GR 2. Neither did they stop the Louisiana field trials in 2004 and 2005 or the IRRI field trials in 2008 and in 2013. Nor did Greenpeace prevent Ashok Singh, Haritha Bollenedi, and others in India from introducing the GR2R event into Swarna in 2009 and again in 2011, or the field trials conducted by Partha Biswas in Bangladesh in 2015, or the two bioconversion studies conducted by Gwangwen Tang. And so on. All these activities and sci-entific advances in Golden Rice technology took place as if Greenpeace, Friends of the Earth, and other professional agitators just didn't exist.

Indeed, the continued progress, improvement, and testing of Golden Rice across the many years of staunch Greenpeace opposition showed, if anything, the actual *ineffectiveness* of their campaigns. All their rhetoric, demonstrations, and other media events constituted a *failure* to block Golden Rice rather than a success.

It is true that Greenpeace's relentless war on GMOs helped to create and sustain a sense of distrust, hostility, and fear of genetically modified foods. And one can argue that there was a consequent spillover effect onto Golden Rice that put it under an equivalent cloud of suspicion. But contributing to a negative perception of a crop was an entirely different matter from committing murder, much less mass murder or crimes against humanity.

It could also be argued that Greenpeace's various activities had created a climate of fear around all GMOs and that this had contributed to the issuance of stringent regulations against their use and development. But those regulations stemmed more from the Precautionary Principle's grip on the minds of lawmakers than from any Greenpeace rhetoric, press releases, position papers, or street demonstrations. The impact, if any, of Greenpeace on the adoption of the Precautionary Principle is at best indirect, tenuous, and unquantifiable.

It is true that Greenpeace had directly impeded, at least temporarily, the development of other genetically engineered foods. In the Philippines, for example, it had tried, and had succeeded for five years, in halting field trials of genetically modified eggplant.

That particular effort had been a circus from beginning to end. On April 26, 2012, members of Greenpeace Manila and some other local NGOs dressed themselves up as GMO "monster crops" and then filed a case asking the Philippine Supreme Court to stop ongoing field trials of insect-resistant *Bt* eggplant, on the novel legal grounds that such open-air testing of this dangerous vegetable violated the constitutional rights of Filipinos to "a balanced and healthful ecology in accord with the rhythm and harmony of nature." Whatever that means.

Incredibly enough, a few weeks later the Philippine Supreme Court granted the petitioners' request and halted the field trials. Greenpeacers were overjoyed by this victory, which they hailed as a major and historic legal milestone. "The Supreme Court decision on Friday sets an important precedent in that it establishes that GE, and in particular GE Bt eggplant, violates the constitutional rights of individuals to a healthy environment," said Greenpeace in a statement. "No other court in the world has upheld such a stance against genetically modified organisms (GMOs). This landmark decision will become subject to national and international legal discourse for years to come."

But the matter did not end there. In 2013 the Philippine FDA issued an opposing judgment saying that GM food was "as safe as and as nutritious as the food derived from conventional crops for direct use as food, feed and for processing." In 2014, a Philippine biotech firm asked the Supreme Court to reverse its decision banning the field trials. And, in that same year, a group of Filipino farmers also asked the court to reverse itself on the grounds that their right to their livelihood as farmers was threatened by being unable to grow the agriculturally successful and quite profitable insect-resistant, genetically modified eggplant.

But, in 2015, the Supreme Court affirmed its original decision. And then, on July 26, 2016, in a major surprise move, that very same court suddenly and unanimously reversed itself, now arguing that "any future threat to a healthful and balanced ecology is more imagined than real."

By 2017, the Philippine government, which had been friendly toward biotechnology since 2001, had approved a total of 88 GMOs for commercial release. That fact gave the lie to Greenpeace's oft-voiced charge that the "real" purpose of Golden Rice was to create a wedge that would open up the Philippines to other biotech crops. No such wedge was needed. For all of its ranting and raving, bombast and bluster, not to mention its monster crop costumes, Greenpeace had been impotent at stopping those 88 GMO approvals.

As for the most part it had also been powerless against the spread of genetically modified crops the world over.* In 2013, when Patrick Moore was starting his Allow campaign, GMOs already accounted for more than 300 million acres of farmland worldwide, grown by over 17 million farmers in more than 25 countries. Further, the vast majority of the increase in planting genetically modified crops was in developing coun-

* In a 2014 journal article, "A Dubious Success: The NGO Campaign against GMOs," Robert Paarlberg shows that NGOs had considerable success in blocking the planting and use of GMOs in Africa, particularly in Zimbabwe and Zambia. But Golden Rice was not intended for those countries.

tries, a process that shows no sign of letting up. Greenpeace had failed to prevent those events, and continues to fail.

■ When Greenpeace complained that Golden Rice still wasn't available for sale "even after more than 20 years of research," the implication was that this was somehow unusual or uniquely damning to the product. And it had been a standard fixture of Greenpeace oratory to claim that Golden Rice scientists had spent "millions" on research and development, as if this, too, were a fault or a weakness that pertained to Golden Rice alone. But were 20 years and millions of dollars spent really excessive or exorbitant when it came to bringing a GMO to market?

In 2011, a British consulting firm was hired by a private company to produce a study of the cost and time involved in the discovery, development, and authorization of new plant varieties created through biotechnology. It collected data from six of the world's major crop biotechnology companies, including BASF Corporation, Bayer CropScience, Dow AgroScience, and Syngenta.

The consulting firm first addressed the question of cost. The biotech corporations provided data concerning the dollar amounts spent in discovering, developing, registering, and obtaining cultivation approval for a single GMO in two countries, plus importation approvals from at least five countries, for a new plant variety that was to be introduced during the years 2008 through 2012. The study found that two of the highest costs incurred in the process were for optimizing the genetic construct to be used in the transformation (an average cost of $28.3 million) and for subsequent introgression of the trait into an "elite," high-yielding plant variety, together with a wide-area field testing program ($28 million). But the expense of dealing with regulatory requirements were the highest of all: an average of $35.1 million, which represented 25.8 percent of the total cost of getting the biotech product to market, which for a single new plant type averaged $136 million.

That may have been a lot of money merely to invent, breed, and bring a new type of plant to the marketplace, but it showed that Golden Rice was not unusual in costing "millions" to develop. The total cost of developing Golden Rice is not known with precision (development is still ongoing), but it is estimated to be approximately $100 million. (That level of spending explains why in most cases only the biggest and richest multinational firms can afford to develop a GMO.)

The lengths of time required to complete each of the several distinct steps involved in creating the new GMO plant variety were also comparably outsized. In the year nearest the date of the British study, 2011, it had taken an average of 2.7 years to optimize the molecular construct, 3.5 years to complete introgression breeding and wide-area testing, and a combined total of 9.3 years to satisfy regulatory science and registration requirements. According to the study report, "Regulatory science, registration and regulatory affairs is the longest single phase in product development and is estimated to account for 25.8% and 36.7% of total cost and time involved respectively."

The total amount of time required for a developer to move from initial trait discovery to final authorization and release into the marketplace was an average of 13.1 years.* "However, there was considerable variation in the responses between companies and between crop species. Overall variability in the responses received varied from a low value of seven years to the comparatively high value of 24 years."

These figures suggest that while the time required to receive authorization to commercialize Golden Rice (if indeed it ever happens) is on the high side of normal, it is not outside the range required for other GMO plant varieties. The figures also show that the single greatest obstacle in developing a GMO for release is compliance with government regulations concerning genetically engineered organisms. It is this, and not any amount of Greenpeace protests, demonstrations, or group chanting, that is responsible for the indefinitely delayed approval and release of Golden Rice.

In the end, this is also the conclusion that Ingo Potrykus himself came to. Initially he thought that Greenpeace was the obstacle and the enemy, but years of experience taught him otherwise. In a 2010 review of the

* Some of the development activities proceeded concurrently, and so the actual development time may be less than the numerical total of the individual time periods mentioned.

"lessons from the 'humanitarian Golden Rice' project," Potrykus said, "Compared to a non-Genetically Engineered (GE) variety, the deployment of Golden Rice has suffered from a delay of at least ten years. The cause of this delay is exclusively GE-regulation."

Exclusively GE-regulation.

To illustrate his contention, Potrykus identified and enumerated the several specific individual requirements that had been established by government directives pertaining to GMOs and the time it took to comply with each of them.

First was the requirement that the selectable marker gene that was part of the molecular construct be deleted from the organism. "This has to be done," Potrykus said, "despite the fact that there is a wealth of scientific literature documenting that the antibiotic marker genes in use have no effect on consumer and environmental safety." Deleting them from the construct required two years of lab work.

Second was the requirement that there be "streamlined integration" of the inserted gene, as opposed to a complicated, random, or messy insertion pattern of the transgene. This is required to prevent possible unintended consequences arising from random insertions that are a common feature of *Agrobacterium*-mediated transformation, which is the means by which Golden Rice had been created. Complying with this demand necessitated repeated experimentation until such a streamlined integration pattern was produced, a process that by itself took another two years.

Third was the requirement for what Potrykus called a "regulatory clean event." This means that the inserted construct had to be a more or less perfect and pristine molecular vehicle, one that did not interfere with the action of neighboring genes, which provided stable expression of the introduced trait at consistent levels and possessed other desired features. "Such 'ideal' events are rarities," Potrykus said. Syngenta had been screening for them as a matter of course, so that all final GR events were "regulatory clean." Still, the screening process took two more years.

The fourth requirement concerned so-called transboundary movements of the Golden Rice seeds, which means shipping them across national borders. "The conditions set up by the Cartagena protocol make exchange of transgenic seed so complicated that it took more than two years to transfer, for example, breeding seed from the Philippines to Vietnam, and one year from USA to India."

Fifth was the requirement that a precise sequence had to be followed in getting the rice from the initial laboratory growth chamber out into the field. There were intermediate steps: growth inside a contained greenhouse and then inside a contained screenhouse. In making these transitions, Potrykus said, "We lost far more than two years."

But even that was not the end. As we have seen, the use of a single lead selection event was not an option that the Golden Rice researchers had embraced voluntarily and on their own. Any breeder would have preferred to work on multiple selection events concurrently. But because it was so hugely expensive to prepare a regulatory dossier for even one event—the cost ran as high as $35 million—the scientists were forced by financial constraints to limit themselves to a single event. Gathering the data required to make such a choice meant another two years of work—and even then the first two events selected were not the correct ones.

And then there was the preparation of the regulatory dossier itself. The dossier was the 1,000-page information package that IRRI presented to the Philippine authorities as evidence that its new plant variety was indeed the miracle of genetic engineering that it claimed. "I will not describe all the hundreds of expensive studies in molecular genetics, biochemistry, nutrition, protein identity, -digestibility, -immunogenicity, gene expression, anti-nutrients, and agronomy required, at publication level quality, for the final regulatory dossier. These requirements keep an entire team of specialists busy for at least four years."

Once a dossier is ready for submission, the actual approval process could take place in less than a year—the fastest part of the whole prolonged enterprise. Still, all these delays added up to a considerable sum, and Potrykus believed that they had cumulatively pushed back the emergence of Golden Rice by at least 10 years. And the cause of that delay, he said, was "*exclusively GE-regulation.*"

■ A delay of ten years or more imposed a substantial health burden on some members of world's poorest populations in the form of years of life and sight lost that otherwise could have been saved by their consumption of Golden Rice. Economists, food scientists, nutritionists, and others took the prospect of Golden Rice seriously enough to put numerical values to some of those burdens. The researchers performed these calculations well before the rice was ready for approval and release, and in one case even before the creation of Golden Rice 2.

In 2004, for example, economists at two German universities published the paper, "Potential Health Benefits of Golden Rice: A Philippine Case Study," in the journal *Food Policy*. Their analysis was admittedly speculative, and its predictive accuracy was limited by a number of uncertainties and unknowns (such as the conversion ratio of beta carotene to retinol). Even so, the authors concluded that Golden Rice "could bring about significant benefits."

Their study utilized the technical metric called the disability-adjusted life year (DALY). This unit of measurement is a means of calculating the overall disease burden on a society, expressed not in terms of the number of lives lost or impaired by disease or disability but rather in terms of *time*, that is, by the number of *years* lost due to ill health, disability, or early death. It is a way of combining losses of various types into a single common unit of measurement that stands for *years of healthy life lost*. As a metric, the disability-adjusted life year was well suited to quantifying the effects of vitamin A deficiency, which includes both disability (blindness) and death.

Using that measurement, the German economists estimated the public health impact of the prototype version of Golden Rice on children and pregnant women in the Philippines. In view of the uncertainties involved in their calculations, the authors produced two separate estimates of the potential health impact of Golden Rice, based on optimistic and pessimistic assumptions about the quantity of beta carotene in the rice kernels, its bioavailability, and the probable extent of program coverage. Even in the most pessimistic scenario, they said, consumption of Golden Rice in the Philippines could save about 15,000 years of healthy life annually. In the optimistic set of assumptions, the rice could save more than 85,000 years of healthy life annually.

"These are remarkable gains," the authors said. But because Golden Rice had been delayed by regulations for more than a decade, an equivalent number of healthy life years had in fact been *lost* every year to VAD.

A later study in 2008, "Genetic Engineering for the Poor: Golden Rice and Public Health in India," by different authors, was based on the higher beta carotene levels produced by Golden Rice 2. India has more than 10 times the population of the Philippines and accordingly would be expected to have significantly higher numbers of years of healthy lives saved by use of Golden Rice. But even so, the actual numbers were surprisingly high. Had Golden Rice 2 been used in India, the authors said, it could have saved 1,382,000 years of healthy life per year. But because it hadn't, all those years of healthy life had been *lost* to disease and death. And that was just in one country.

THE APPROVALS

The first countries to formally approve Golden Rice as a food for human consumption were Australia and New Zealand, which jointly accepted the product for such use in December 2017. This was both surprising and strange. First of all, it was surprising to see Golden Rice finally achieve the status of an approved food product anywhere in the world after almost 20 years of unremitting delays, disappointments, and failed experiments during the course of otherwise continuous product development. What was strange about it was that in neither of these countries was vitamin A deficiency a problem. Why, then, were they the first to approve it? For that matter, why did the International Rice Research Institute even bother to file an application for commercialization of the rice in those two nations?

An official explanation was given by Food Standards Australia New Zealand (FSANZ), the government agency that regulated and enforced food standards for both countries. In its December 20, 2017, *Approval Report—Application A1138*, FSANZ stated that "rice containing the GR2E event is not intended for commercialization in Australia or New Zealand, i.e., either for growing or intentional sale in the food supply. The Applicant has however applied for food approval because it is possible the rice could inadvertently enter the food supply via exports from countries that may supply significant quantities of milled rice to Australia or New Zealand." In other words, if small amounts of the rice were accidently present in shipments of conventional imported rice, no trade issues would arise.

That explanation was itself somewhat incongruous given that Golden Rice proponents had always claimed that accidental admixture or commingling of ordinary rice and Golden Rice would be unlikely because of the color of the grains. Still, accidents could happen.

Adrian Dubock, however, had a different take on the matter, focusing on the possible role of "accidents." Writing in May 2018, he said, "The opponents are so insistent in their views that it wouldn't surprise me to find an inclusion of Golden Rice in a shipment of white rice which was not accidental." That is, put there by an act of sabotage—or in Greenpeace-speak, "ecotage."

Given the history of anti-GMO activities to date—actions that included many instances of field-test vandalism, including the destruction of the GR2R test plot at Pili in the Philippines in 2013—the possibility of sabotage was not implausible. Ingo Potrykus's bomb-proof greenhouse was a constant reminder that activist violence was an ever-present threat.

A different explanation offered by others was that IRRI had sent its application to countries that were thought likely to grant approval, in an effort to start a domino-effect wave of further approvals by other countries. But that explanation was itself suspect, as neither Australia nor New Zealand were particularly GMO friendly.

Whatever the true reason was, the FSANZ approvals were not big news, nothing like the *Time* magazine cover story about Golden Rice and Potrykus back in the year 2000. The approvals were reported mainly in specialized scientific print media and on biotech websites such as that run by the Genetic Literacy Project. However, the approvals caused a veritable explosion of discontent, an eruption of outrage, an upheaval of fury, among anti-GMO and Golden Rice opposition groups. Indeed, the number, stridency, and falsity of the many claims they made about the approvals, as well as the application document that led to them, were striking to behold.

The FSANZ approvals were announced at the end of December 2017. At the beginning of the following year, a group called GE-Free NZ was already asking the New Zealand minister of food safety to reconsider its decision to approve Golden Rice. The group spoke of a "total absence of data" regarding the safety of the rice, a claim that to anyone who had actually read (or even skimmed) the 1,000-page application document was preposterous. The group was soon threatening to take legal action against FSANZ "for failing in its mission to protect the health of the public," even though the approval report explicitly stated that GR2E was *not intended to be used in the Australian or New Zealand food supplies.*

On February 5, 2018, a group in Germany, Testbiotech, claimed that "no toxicological studies were performed with the rice." This was also flatly untrue.

Allison Wilson, a longtime GMO critic writing on the *Independent Science News* web page, said of the approvals, "These are in essence fake approvals as they are in countries that do not intend to grow or eat GR2-E. It appears these countries have approved GR2-E to please the GMO industry and to put pressure on countries like the Philippines to approve GR2-E for cultivation, even though a final GR2-E product is not yet available and no GR2-E line has undergone adequate testing for efficacy and safety."

But in fact both the PhilRice/IRRI original application package and FSANZ's approval report (which itself ran to more than 100 pages) addressed safety, toxicology, and allergenicity issues at length and in depth. The application package included separate safety reviews of each of the three proteins that were expressed in the rice: the *Zm*PSY1 protein from maize, the PMI mannose isomerase protein that had been used as a selectable marker, and the CRTI carotene desaturase protein.

The application noted that the PMI mannose protein had already been used successfully in a wide range of genetically engineered maize lines and posed no safety risks. As for the *Zm*PSY1 and the CRTI proteins, both were found to be rapidly and fully digested in vitro by simulated gastric juices. In addition, the enzymatic activity of both proteins was completely destroyed by heat, at levels well below cooking temperature.

Because the CRTI protein came from a nonfood source, the bacterium *Erwinia uredovora* (recently renamed *Pantoea ananatis*), IRRI scientists had conducted an acute oral toxicity study of the protein in mice. All the animals had done well, gaining weight during the study and suffering no toxic or other adverse health effects. For these and other reasons, the PhilRice and IRRI scientists concluded that all three of the proteins that they had examined, screened, and tested were unlikely to be toxic or allergenic to humans or animals.

In discussing any new or proposed genetically modified food, GMO critics have predictably raised the issue of possible unintended consequences resulting from the product's introduction and use. And so, in the application dossier, the PhilRice and IRRI scientists addressed this problem as well. They undertook an extensive compositional analysis of GR2E rice samples and compared the results to the presence, absence, or levels of expression of 69 separate compositional components in control, unmodified rice that had been grown at four different sites in the Philippines in 2015, and again in 2016, and concluded from these analyses:

> The only statistically significant difference observed from the multi-year combined-site analysis was for stearic (C18:0) acid, a minor fatty acid component. . . . With the exception of β-carotene and related carotenoids, the compositional parameters measured in samples of GR2E rice, including stearic acid, were within or similar to the range of natural variability of those components in conventional rice varieties with a history of safe consumption. Overall, no consistent patterns emerged to suggest that biologically meaningful changes in composition or nutritive value of the grain or straw had occurred as an unexpected, unintended consequence of the genetic modification.

Since there were no significant changes in the material makeup of the two rices other than the presence of stearic acid, "which was approximately 6.5 percent higher for GR2E rice," and the deliberate and intended presence of beta carotene, there were unlikely to be unintended effects from the adoption and use of Golden Rice.

The scientists at Food Standards Australia New Zealand received the PhilRice/IRRI application in November 2016. It took the scientists about a year to evaluate all the data, and in the end they produced their own 103-page document in which they stated and explained their findings and their reasons for approving Golden Rice GR2E: "No potential public health and safety concerns have been identified. Based on the data provided in the present Application, and other available information, food derived from line GR2E is considered to be as safe for human consumption as food derived from conventional rice cultivars."

And that was that.

In 2017, Health Canada also received essentially the same submission to assess the safety of Golden Rice for possible use in Canada, although the government also stated that "this product is not intended to be sold in Canada at this time." It took the relevant Canadian scientists about a year to complete their own separate review of the materials package. And, in March 2018, Health Canada approved Golden Rice for sale in that country, saying, "Following this assessment, it was determined that the changes made in this rice variety did not pose a greater risk to human health than rice varieties currently available on the Canadian market. In addition, Health Canada also concluded that GR2E would have no impact on allergies, and that there were no differences in the nutritional value of GR2E compared to other traditional rice varieties available for consumption except for increased levels of provitamin A."

What these approvals meant was that the health authorities, nutritional scientists, toxicologists, and experts in molecular biology and chemistry in three of the world's most technologically advanced and scientifically literate countries had examined all the data concerning what was arguably one of the most fully studied and exhaustively documented foods in the history of human nutrition, and pronounced it as safe to eat as conventional rice for their very own citizens.

■ But none of this—neither the approvals themselves nor the rafts of data on which they were based—made the slightest difference to those under the spell of what amounted to anti–Golden Rice rapture fever. In April 2018, the Stop Golden Rice! Network held a three-day international conference at the Madison 101 Hotel and Tower at Quezon City, just north of Manila, in the Philippines. The theme of the conference, as announced on posters was "Amplify our call to end corporate control over agriculture. Stop Golden Rice! Defend farmers' rights and food sovereignty." The conference was attended by members of more than two dozen Golden Rice opposition groups, including MASIPAG,* which was largely responsible for what one of their press releases proudly described as "the historic uprooting of the Golden Rice field trials in 2013."

On the second day of the conference, more than 100 attendees staged a protest at the Philippine Department of Agriculture calling for the agency "to immediately scrap the application for open field test and direct use of Golden Rice in the Philippines." The protestors also called for "the immediate closure of the said institute [IRRI]" on the grounds that "farmers have become more indebted and poorer under IRRI's Green Revolution program." This was incredible since it was precisely the varieties developed by Peter Jennings and Gurdev Khush at IRRI that constituted the bulk of the rice grown and actually consumed in the Philippines.

Two days later, on April 20, the Stop Golden Rice! Network launched a petition to halt the commercialization of Golden Rice on the grounds that it posed a threat to the health of women and children. In a press release, Chito Medina, identified as an environmental scientist, was quoted as saying, "What studies are available right now point out that Golden Rice will be more of a risk than a benefit. In India, rice plants that have been bred with the Golden Rice showed abnormal growth that might affect the farmers' cultivation." Medina did not mention that the rice in question was the GR2R Swarna variety that had been long since abandoned by all developers and that the rice now awaiting approval in the Philippines was, rather, the GR2E variety that thrived and grew normally.

But neither respect for facts nor fairness played any significant role in motivating the activists. What their rhetoric, demands, and actions really demonstrated was that no amount of evidence, data, test results, or reasoned argument was going to affect their dogmatic opposition to the "corporate greed" that was allegedly behind Golden Rice, which they proclaimed was a "political crop," a "toxic product," and "a death reaper to our local rice sector."

---

* Magsasaka at Siyentipiko para sa Pag-unlad ng Agrikultura (Farmer-Scientist Partnership for Development).

■ And then, in May 2018, came the most surprising news of all: the United States approved Golden Rice. The US government had already recognized the value of the product when in 2015 the White House Office of Science and Technology Policy together with the US Patents and Trademark Office bestowed its Patents for Humanity award upon Golden Rice. The award recognizes patent holders working to improve global health and living standards for the populations of developing nations. On April 20, 2015, Adrian Dubock formally accepted the award on behalf of the inventors, Ingo Potrykus and Peter Beyer, during a ceremony at the White House.

The official FDA approval was indeed a historic development in the annals of Golden Rice, but other than for accounts published in science journals and on pro- and anti-GMO websites, the approval failed to generate the least ripple of news or interest.

There were in fact some good reasons for this. To begin with, the approval letter was buried deep inside the FDA's bureaucratic announcement apparatus, appearing as an item in the "Biotechnology Consultations on Food from GE Plant Varieties" section of the FDA website. Further, the letter itself, which was signed by Dennis M. Keefe, director of the FDA's Center for Food Safety and Applied Nutrition, was peculiar in several respects. For one thing it was undated (though an external reference showed that it had been produced on May 24, 2018). For another, it was addressed to Donald J. MacKenzie, of the IRRI office in Manila, although MacKenzie had since moved to the Donald Danforth Plant Science Center in St. Louis, Missouri.

A further peculiarity was that the two-page letter was little more than a summary of the contents of the submission that the FDA had received from IRRI on November 14, 2016. At the end of the summary, Keefe concluded, "Based on the information IRRI has presented to FDA, we have no further questions concerning human or animal food derived from GR2E rice at this time."

If that sentence constituted the FDA "approval," it was scarcely a ringing endorsement. However, language that more closely resembled an endorsement was contained in another document, equally well buried on the FDA web page, namely, "Biotechnology Notification File No. 000158," by Carrie McMahon, FDA consumer safety officer.

This eight-page notification file laid out the FDA's position in greater detail, saying in its formal conclusion that "FDA evaluated IRRI's submission to determine whether GR2E rice raises any safety or regulatory issues with respect to its uses in human or animal food. Based on the information provided by the company and other information available to the agency, FDA did not identify any safety or regulatory issues under the FD&C [Food, Drug and Cosmetic] Act that would require further evaluation at this time."

For the FDA to publicly state that its scientists had no safety or regulatory issues with Golden Rice GR2E was itself an important announcement, although the way in which it was couched still made it into a rather lukewarm endorsement of the product. In addition, a qualification in Dennis M. Keefe's May 24 letter seemed to undercut the value of Golden Rice as a food: "Although the concentration of ß-carotene in GR2E rice is too low to warrant a nutrient content claim, the ß-carotene in GR2E rice results in grain that is yellow-golden in color."

Too low to warrant a nutrient content claim.

This seemed to dismiss the value of Golden Rice in the very act of clearing it for use as a food. "Too low" according to what criterion? What kind of a nutrient content claim? None of this was explained anywhere in Dr. Keefe's letter.

But to Golden Rice critics, Keefe's elliptical statement was like manna from heaven. On June 3, 2018, Allison Wilson and Jonathan Latham wrote on the *Independent Science News* web page a story they headlined, "GMO Golden Rice Offers No Nutritional Benefits Says FDA."

No nutritional benefits. Plainly, the FDA did not say quite that.

And so on the following day, June 4, in an attempt to clarify the issue, this author wrote to Dennis M. Keefe, referenced the *Independent Science News* headline, and asked, "Is this an accurate interpretation of your remark? And if not, could you clarify the sense in which the rice has some nutritional value, however low it might be?"

Three days later, on June 7, I received the FDA's reply for the record, quoted below in full (all emphasis as in original):

It is unfortunate that the statement you reference in our letter responding to BNF 158 has been misconstrued to suggest that there would be no value of the pro-vitamin A in golden rice **for its use in the countries where it is intended for distribution.**

Our statement applies only to labeling considerations in the **United States,** in that golden rice contains insufficient pro-vitamin A to warrant differential labeling for nutrient content based on the low levels of rice consumption in the U.S. Requirements for nutrient content claims on labels in the United States take two factors into account, the amount of the nutrient needed as well as its concentration in the food and the typical or average level of that food consumed in the U.S. For the rice to be labeled in the United States with a claim **containing pro-vitamin A,** our regulations stipulate that the food must contain 10–19 percent of the RDI [reference daily intake] or DRV [dietary reference value] for the substance per reference amount customarily consumed (essentially a measure of consumption).

Additionally, U.S. consumers eat rice at very low levels compared to consumers in the specific Asian countries with vitamin A deficiency for which golden rice was developed. IRRI reports that consumption of rice by children in Bangladesh is 12.5 g / kg body weight / day (compared to about 0.5 g / kg bw / d for U.S. consumers). Rice is the major staple in those countries and levels of rice consumption are many-fold higher than they are in the U.S. While a U.S. consumer would be unlikely to eat enough of the rice to achieve that value (10–19% of the NDI [new dietary ingredient] or RDA [recommended dietary allowance]), that does not mean that the level of consumption of golden rice in the targeted countries would be insufficient to accomplish the intended effect of supplementing their very low consumption of vitamin A–containing foods. Consuming rice containing the levels of pro-vitamin A in GR2E rice as a staple of the diet could have a significant public health impact in populations that suffer from vitamin A deficiency.*

Here the FDA stated in as many words that Golden Rice GR2E *could have a significant public health impact in populations that suffer from vitamin A deficiency*. Which is exactly what Golden Rice proponents had been saying all along. For the US Food and Drug Administration to affirm this view was a major victory for Golden Rice.†

What remained was for the rice to be approved by those nations that actually needed it. But this process has dragged on and on, even in the Philippines, the very country whose scientists had produced the original application for approval.

---

* Marianna Naum, PhD, team lead, Strategic Communications, Office of Foods and Veterinary Medicine, U.S. Food and Drug Administration, email to Ed Regis (cc: Dennis M. Keefe), June 7, 2018.

† On June 7, 2018, Ed Regis posted a complete copy of the FDA's June 7 email as a comment on the *Independent Science News* web page. When asked to retract their story headlined "GMO Golden Rice Offers No Nutritional Benefits Says FDA," the authors declined. The false headline remains on their web page to this day.

EPILOGUE

THE PROACTIONARY PRINCIPLE

An aphorism attributed to Friedrich von Schiller, the German poet, says, "He that is overcautious will accomplish little." But in fact that is woefully understating the case. Overcautiousness can instead destroy much, in the form of opportunities lost, benefits forgone, and costs incurred by avoiding apparent threats that in the end turn out to be imaginary. The Precautionary Principle was a classic example of the price paid by overcaution, and the case of Golden Rice made it abundantly clear that, when it came to evaluating the benefits and risks of a valuable new technology, adherence to the Precautionary Principle did far more harm than good. It is arguable, therefore, that the principle ought to be replaced with a different one, by an alternative approach that is designed to foster innovation and allow us to reap the benefits offered by new products and advanced technologies. But the question was, replaced with what?

Golden Rice is unique as a GMO because it is the first example of a plant that was genetically engineered not for the usual purpose of making a given crop resistant to herbicides, pesticides, or pathogens. Those qualities, as valuable as they are in lowering the costs of producing maize, soybeans, or whatever else, mainly tend to benefit farmers. Golden Rice, by contrast, is the first GMO crop designed from the start to benefit the *user*: the consumer, the individual who eats the product, ingests a needed micronutrient, and becomes healthier in the process. Golden Rice is unique because it is the first plant that was intentionally engineered to have that attribute, but it would not be the last, for other such crops were already being developed. The golden banana, for example.

In 2017, the golden banana was going through what amounted to a video replay, rebroadcast, or live reenactment of the Golden Rice saga. The story of its creation did not produce the same media sensation as Golden Rice had back in the year 2000, but the July 7, 2017, issue of the online journal *Science Daily* featured the headline, "Golden Bananas High in Pro-vitamin A Developed." And there on the page was pictured another bearded, white-haired man surrounded by a sea of green foliage—the broad green leaves and thick stalks of banana plants. The image was almost a duplicate, intentional or not, of the *Time* magazine cover photo of Ingo Potrykus against a backdrop of rice greenery, but in this new case the caption read, "Distinguished Professor James Dale has led a team of scientists who have modified bananas to make them rich in pro-vitamin A. This humanitarian project will improve the nutrition of people in Uganda."

Professor Dale and his team did their work at Queensland University of Technology in Brisbane, Australia. Their project was similar to that of Golden Rice in several respects. First, it was an attempt to genetically engineer the beta carotene trait into a crop, the East African highland cooking banana that is a staple food among the poor in a region where vitamin A deficiency is at high levels, in this case, Uganda. Second, to achieve this goal the Australian scientists employed the same gene that the Syngenta team had utilized to create Golden Rice 2: the gene for maize phytoene synthase, *ZmPsy*. The Australian scientists inserted that gene into a banana variety, the familiar Cavendish cultivar, which was different from the intended target cultivar. This also paralleled the way Golden Rice was invented, first in one rice variety (Taipei 309), with the beta carotene trait later transferred into one or more other varieties. Further, the fruit that resulted was of a different color from the original: no longer cream colored, it was now of a golden-orange hue. That is the telltale sign of beta carotene, and it is a change that the fruit's intended recipients would have to get used to, just as consumers of white rice would have to get used to the yellow color of Golden Rice.

There was one major and significant difference, however, between the Golden Rice development process and that of the golden banana.

Whereas Potrykus and company had been forced to choose a single lead selection event on the basis of molecular and greenhouse data, James Dale and his team had their developmental road virtually paved with gold. They had "applied for and were granted approval from the Australian regulator to take nearly 2000 independent events to the field with minimal prior characterization," according to the project's published history. This was a boon to their research program because it took at least 3.5 years from the point of the initial transformation to the time of harvesting the first fruit in the field.

"Our approach therefore was to test multiple transgenes and multiple promoters and different combinations thereof in parallel. . . . This approach was successful in identifying two promoters and one transgene that would result in target levels of pro-vitamin A elevation without any obvious phenotypic abnormalities."

Without question, that was the fast-track, accelerated method of getting a GMO optimized—a luxury that Potrykus, Beyer, and colleagues could only dream about.

But there were still some regulatory hurdles ahead for Dale and his crew. After their success in creating the golden Cavendish banana, the researchers sent the genes in test tubes to breeders in Uganda for crossing into the local East African highland variety and for field trials—at least *confined* field trials. Uganda had established a Biotechnology and Biosafety Policy in 2008, but the policy governed GMOs only through the confined field trials stage. The Ugandan Parliament had several times debated a new law with wider scope that would cover the whole development process through to the end, but thus far without resolution.

Several years later, by 2014, the country had still not established an appropriate regulatory framework concerning the release and commercialization of genetically modified organisms. Uganda was a signatory to the Cartagena Protocol, however, and so it was bound by the Precautionary Principle, meaning that there was no guarantee that legislators, if and when they enacted the necessary laws, would adopt a regulatory framework that was favorable to further GMO development.

But in October 2017, at long last, the Parliament passed the National

Biosafety Act of 2017. Its provisions were quite liberal and would ultimately allow farmers and consumers free and open access to the benefits
of genetically modified organisms. The country's crop scientists and others regarded passage of the bill as a monumental victory for science. The
act would become effective as soon as the country's president, Yoweri Museveni, signed the bill into law, which he was confidently expected to do.

But he didn't. On December 21, 2017, President Museveni returned
the bill to Parliament, citing several concerns about GMOs, including
the possibility of cross-contamination of genetically modified seeds
with non-GMO, native species. "To be on the safe side," he said in a letter to Parliament, "GMO seeds should never be randomly mixed with
our indigenous seeds just in case they turn out to have a problem."

To be on the safe side. Just in case. The Precautionary Principle had
struck again.

■ If the Precautionary Principle is about any one thing, it is about safety.
It is about risk avoidance, about averting unproven, negligible, improbable, hypothetical, and even "unknown" risks—*especially* unknown risks.
And if the principle has one major defect, it is that it ignores or mini-
mizes the benefits that might counterbalance, outweigh, or even nullify
the magnitude, if any, of those risks.

In February 2004, a group of technology theorists, cultural thinkers,
and futurists convened on the internet for a Vital Progress Summit
meeting, a virtual gathering of the minds similar to what we would now
call a webinar. The meeting was sponsored by the Extropy Institute,
which was devoted to ideas once regarded as so advanced as to be science fictional, such as life prolongation, that have since become quite
mainstream even among the most orthodox scientists. Since maximiz-
ing the human lifespan might require the use of new and experimental
technologies and methods, members of the Vital Progress Summit
group gave careful consideration to the question of what principle or
principles would best promote and protect the necessary technological
discoveries, breakthroughs, and inventions.

The list of virtual attendees included Aubrey de Grey, a University of

Cambridge geneticist; Marvin Minsky, an MIT computer scientist; Ray Kurzweil, author and futurist; Lee Silver, molecular biologist at Princeton; Roy Walford, of the UCLA School of Medicine; and Max More, an independent scholar with a PhD in philosophy from the University of Southern California. Their discussions produced a great wealth of ideas, key among which was that the Precautionary Principle is not well suited to fostering the kinds of innovation and experimentation needed to make fundamental technological progress and indeed in many ways it actually sabotages scientific advance. An alternative principle, or set of them, should be formulated and put in its place.

At the end of the virtual summit meeting, Max More combined several of the strains of thought that had arisen and fashioned them into a coherent doctrine that he set forth as the "Proactionary Principle," in conscious and deliberate opposition to the Precautionary Principle. The Proactionary Principle is different from the Precautionary Principle in several ways, the most fundamental being that it is characterized, first of all, by multiplicity. This is an attribute that makes it more congruent with the multifaceted nature of technology development.

The invention of a new technology is not a simple, linear, streamlined process with only a single, self-contained, and circumscribed effect. The development and operation of any new technology constitutes an interference into an already existing, complex system of causes and effects, and therefore it invariably has an indefinite number of intended and unintended consequences. Those consequences, both good and bad, are multidimensional: they have a certain magnitude, likelihood, value, and proximity in time, among other properties. To evaluate a new technology, therefore, requires more than a single, simplistic, and unitary governing law. It requires a *system*, a decision procedure, one that is composed of several interlocking rules.

It is important, first of all, that the decision process be *comprehensive*. That is, it must take into account *all* of the consequences of a given innovation—positive and negative. It is shortsighted to focus on risks alone, as if benefits don't exist or don't matter. "Estimate the opportunities lost by abandoning a technology," Max More wrote in an account of the principle. "When making these estimates, . . . carefully consider not only concentrated and immediate effects, but also widely distributed and follow-on effects."

The proactive approach also *prioritizes consequences*. This means that it gives precedence to known, proven, and more certain risks and benefits as opposed to those that are merely hypothetical, remotely possible, improbable, or unknown. Priority must be given to immediate benefits over more distant threats or fears. Positive effects that are more lasting or persistent are preferable to those that are fleeting or more transient.

In addition, the proactionary approach embraces the idea of *proportionality*. One ought to apportion precautionary measures to the magnitude and likelihood of possible risks, which should be assessed according to the degree to which they might actually materialize as real and genuine threats.

Forgoing genuine and substantial benefits to avoid threats that are largely notional is not a rational course of action. "Consider restrictive measures only if the potential negative impact of an activity has both significant probability and severity," Max More said. "In such cases, if the activity also generates benefits, discount the impacts according to the feasibility of adapting to the adverse effects. If measures to limit technological advance do appear justified, ensure that the extent of those measures is proportionate to the extent of the probable effects."

Finally, the proactionary approach includes the idea of *symmetrical treatment*, meaning that technological risks must be treated on the same basis as natural risks. "Avoid underweighting natural risks and overweighting human-technological risks."

In sum, the Proactionary Principle constitutes a set of rules and graduated weighting scales that, taken together, have the effect of rationally and systematically subordinating improbable risks to probable benefits. Its multipart, multidimensional nature makes it better suited as a decision procedure to evaluate the prospective effects of interventions into natural or artificial systems—such as the introduction of a GMO into an ecosystem—than a one-dimensional dictate such as the Precautionary

Principle, which myopically focuses on the single and narrow goal of avoiding even the remotest possible danger.

The proactionary approach would in particular have accelerated the development of Golden Rice. Vitamin A deficiency is a known, proven, and genuine public health problem that affects millions, and this is of inestimably greater moment than the speculative and unknown negative consequences hypothesized by Golden Rice opponents. The effects of withholding, delaying, or retarding Golden Rice development through overcautious regulation imposed unconscionable costs in terms of years of sight and lives lost, and these costs were, in the case of death, irreversible.

In retrospect, of course, all this was quite obvious. As Ingo Potrykus has said, "Regulation must be revolutionized." Unfortunately, the revolution has not yet come.

A crime? No. A tragedy? Yes.

ACKNOWLEDGMENTS

This book could not have been written without the substantial help of the three Golden Rice principals: Ingo Potrykus, Peter Beyer, and Adrian Dubock. I am deeply grateful for their trust and cooperation, and especially for the wealth of information, documents, photographs, and miscellaneous data points provided to me by Dubock.

In addition, I would like to thank the following individuals for their assistance: Bruce M. Chassy, Nina Federoff, Mike Grusak, Gurdev Khush, Belinda Martineau, Alan McHughen, Max More, Marianna Naum, Catherine Price, Richard D. Semba, Aron Silverstone, Asok Singh, Gary Toenniessen, and Keith P. West Jr.

APPENDIX

L'AFFAIRE SCHUBERT

N o discussion of the safety of Golden Rice, or of GMO foods in general, would be complete without including the views of David Schubert, a Salk Institute professor and head of the institute's Cellular Neurobiology Laboratory. In October 2002, Schubert published a one-page paper in *Nature Biotechnology*, "A Different Perspective on GM Food," in which he argued that "GM food is not a safe option." Schubert laid out a three-point argument. His first point was that the introduction of the same gene into two different cell types can produce "two very distinct protein molecules." His second point was that the introduction of the new gene "significantly changes overall gene expression and therefore the phenotype of the recipient cell." And the third was that introducing enzymatic pathways that synthesize small molecules, "such as vitamins," could interact with preexistent pathways "to produce novel molecules."

"The potential consequence of all of these perturbations," Schubert said, "could be the biosynthesis of molecules that are toxic, allergenic, or carcinogenic."

Worse still, some of these genetically perturbed plants, such as "a GM plant making vitamin A," might produce substances that could even cause "abnormal embryonic development," that is, birth defects. A GMO that produced vitamin A, in other words, might be the plant equivalent of thalidomide. Plainly, this is something we can do without.

These were serious charges, and they merited a serious reply. And in the December 2002 issue of *Nature Biotechnology* (just two months after Schubert's piece had appeared), they got one. The reply was remarkable on several counts. For starters, the response, which was also just one page long, was the joint product of *18* individual authors, 4 of whom were colleagues of Schubert's at the Salk Institute. More important was the sweeping, final, and apparently unanswerable nature of their criticisms.

The authors recapitulated Schubert's three points and then added, "The fatal flaw of this argument is that each of these three scenarios raised can and do occur in nature during the course of the random 'natural' gene mutations and rearrangements that drive evolution." In other words, each of Schubert's three points was valid, but there was nothing about them that pertained uniquely to GMOs as opposed to "natural" plants and animals.

"In the world of plant breeding," the 18 authors continued, "new genes and different alleles are constantly being shuffled and reshuffled in an endless array of combinations. In the end, traditional breeding practices all have the same, if not greater, potential ability as a transgene to produce different proteins in different cell types, or produce proteins that will react with other substances in a cell to result in 'the biosynthesis of molecules that are toxic, allergenic, or carcinogenic.' "

The authors paid special attention to Schubert's claim that a genetically modified plant that made vitamin A might produce substances that caused birth defects. To this they replied, first, that there is no such plant, and that Golden Rice had been engineered to produce not vitamin A but a vitamin A *precursor*, beta carotene. Second, "all green plants synthesized its precursors," meaning that there is nothing special about Golden Rice that makes it any more likely than any other green plant to cause birth defects.

The authors made several other equally damaging, and even damning, points and closed by bemoaning that fact that Schubert's error-ridden piece had ever been published in the pages of *Nature Biotechnology*.

But David Schubert was not yet done with Golden Rice. Six years later, in 2008, he came at the subject again, in "The Problem with Nutritionally Enhanced Plants," which was published not in *Nature Biotechnology* but in the lesser-known *Journal of Medicinal Food*. His argument this time was more narrowly focused. He correctly identified Golden Rice's end product as beta carotene rather than vitamin A. He then observed that, "upon ingestion by animals, ß-carotene is cleaved in half by a dioxygenase to generate retinal for use in the retinal cycle."

Problems arise, he said, when retinal is oxidized to retinoic acid (RA), a so-called apocarotenoid, a compound chemically derived from carotenoids. Retinoic acid is important because it plays a role in embryonic development. "Excess RA or RA derivatives are exceedingly dangerous," Schubert said, "particularly to infants and during pregnancy."

Schubert once again suggested, therefore, that Golden Rice could cause birth defects. He presented no actual evidence that Golden Rice does, or even that it can in fact produce "excess" retinoic acid levels; he left it as a mere possibility. For that reason the case he made was quite speculative, to say the least.

Taken by itself, such an argument would hardly be worth answering. Coincidentally, however, there is a separate body of evidence that very high doses of beta carotene (30 mg/day) are associated with increased levels of lung cancer among smokers and in those who have been exposed to asbestos. Further, animal studies suggest that beta carotene oxidation products, including apocarotenoids such as retinoic acid and others, are involved in these cancers.

In the eyes of Peter Beyer at Freiburg and Parminder Virk at IRRI, the possibility that apocarotenoids are carcinogenic would call into question the safety of Golden Rice and was therefore an issue that had to be addressed. "To deal with it on a data-based manner forced us through a year-long study on carotenoid degradation products," Beyer said.[*]

So Beyer and Virk, together with colleagues at Freiburg and elsewhere, undertook a comprehensive review of the health effects of foods containing beta carotene. Aiming at an Asian context, the team analyzed 102 different plant foods commonly eaten in the Philippines. Some of these foods had exotic names unfamiliar to Westerners, such as sili leaves, camote tops, mustasa, pechay, and kangkong, but the full list of food items they evaluated also included Golden Rice as well as yellow and orange soft drinks whose color is provided by beta carotene. (The

[*] Peter Beyer, email to Ed Regis, April 12, 2018.

scientists also discovered during the course of their research that a certain apocarotenoid [ß-apo-8′ carotenal] was used worldwide as a food colorant and that the European Safe Food Authority had even established an acceptable daily intake of the substance.)

The team's analysis showed that the apocarotenoids produced by the foods they studied existed in such minute amounts that unrealistically high quantities of them—for example, 7 pounds of carrots per day—would have to be consumed to yield the same level of apocarotenoid provided by 30-milligram daily megadoses of beta carotene.

The scientists published their results in the *Journal of Agricultural and Food Chemistry* in 2017. "Our analysis and quantification of ß-carotene-derived cleavage products across biofortified and nonbiofortified crop plant tissues combined with the calculation of potential exposure document no reason for concern," the authors wrote. "We thus cannot see reasons for banning the orange carrot (that became provitamin A biofortified through selection in the 14th century) or any other crop plant high in beta carotene for the avoidance of 'dark sides' in populations suffering from deficiency."

None of the foods that the scientists reviewed was found to be carcinogenic, much less were they the plant equivalent of thalidomide.

Thus ended *l'affaire* Schubert.

BIBLIOGRAPHY

PREFACE

Cartagena Protocol on Biosafety to the Convention on Biological Diversity. 2003. http://bch.cbd.int/protocol/text/.

Church, George. "The Augmented Human Being: A Conversation with George Church." Edge.org, March 30, 2016 [Recorded 2015]. https://www.edge.org/conversation/george_church-the-augmented-human-being.

Greenpeace. "Genetically Engineered 'Golden Rice' Is Fools Gold." Press release, February 9, 2001. http://www.greenpeace.org/seasia/ph/press/releases/genetically-engineered-golden-2/.

Nash, J. Madeline. "This Rice Could Save a Million Kids a Year." *Time*, July 31, 2000. http://content.time.com/time/magazine/article/0,9171,997586,00.html.

Potrykus, Ingo. "GMO-Technology and Malnutrition: Public Sector Responsibility and Failure." *Electronic Journal of Biotechnology*, North America, 8 (2012). http://www.ejbiotechnology.info/index.php/ejbiotechnology/article/view/1116/1498.

———. "Lessons from the 'Humanitarian Golden Rice' Project: Regulation Prevents Development of Public Good Genetically Engineered Crop Products." *New Biotechnology* 27 (2010): 466–72. https://doi.org/10.1016/j.nbt.2010.07.012.

Shiva, Vandana. "The 'Golden Rice' Hoax." n.d. http://online.sfsu.edu/rone/GEessays/goldenricehoax.html.

CHAPTER 1 CHILD KILLER

Brooks, Sally. *Rice Biofortification: Lessons for Global Science and Development.* London: Earthscan, 2010.

Dowling, John E., and George Wald. "Vitamin A Deficiency and Night Blindness." *PNAS* 44 (July 15, 1958): 648–61. https://www.ncbi.nlm.nih.gov/pmc/articles/PMC528639/.

Price, Catherine. *Vitamania: Our Obsessive Quest for Nutritional Perfection*. New York: Penguin, 2015.

Reddy, V. "History of the International Vitamin A Consultative Group 1975–2000." *American Society for Nutritional Sciences* (2002): 2852S–56S.

Semba, Richard D. "The Discovery of the Vitamins." *International Journal for Vitamin and Nutrition Research* 82 (2012): 310–15. https://doi.org/10.1024/0300 -9831/a000124.

———. "On the 'Discovery' of Vitamin A." *Annals of Nutrition & Metabolism* 61 (2012): 192–98. https://doi.org/10.1159/000343124.

———. *The Vitamin A Story: Lifting the Shadow of Death*. Basel: Karger, 2012.

Sommer, Alfred. "Increased Mortality in Children with Mild Vitamin A Deficiency." *Lancet* 322 (September 10, 1983): 585–88. https://doi.org/10.1016/S0140 -6736(83)90677-3.

———. "Vitamin A Deficiency and Clinical Disease: An Historical Overview." *Journal of Nutrition* 138 (2008): 1835–39. https://doi.org/10.1093/jn/138.10.1835.

Sommer, Alfred, and Keith P. West Jr. *Vitamin A Deficiency: Health, Survival, and Vision*. New York: Oxford University Press, 1996.

World Health Organization. "Global Prevalence of Vitamin A Deficiency in Population at Risk: 1995–2005." http://www.who.int/vmnis/database/vitamina/x/en/.

Zhong, Ming, et al. "Retina, Retinol, Retinal and the Natural History of Vitamin A as a Light Sensor." *Nutrients* 4 (2012): 2069–96. https://doi.org/10.3390/nu412 2069.

CHAPTER 2 **PROOF OF CONCEPT**

Buckner, Brent, et al. "Cloning of the *y1* Locus of Maize, a Gene Involved in the Biosynthesis of Carotenoids." *Plant Cell* 2 (September 1990): 867–76. http://www.plantcell.org/content/2/9/867.

Buckner, Brent, et al. "The *y1* Gene of Maize Codes for Phytoene Synthase." *Genetics* 143 (1996): 479–88. https://www.ncbi.nlm.nih.gov/pmc/articles/PMC120 7279/.

Burkhardt, P. K., et al. "Transgenic Rice (*Oryza sativa*) Endosperm Expressing Daffodil (*Narcissus pseudonarcissus*) Phytoene Synthase Accumulates Phytoene, a Key Intermediate of Provitamin A Biosynthesis." *Plant Journal* 11 (1997): 1071–78. https://doi.org/10.1046/j.1365–313X.1997.11051071.x.

Christensen, Jon. "Scientist at Work: Ingo Potrykus; Golden Rice in a Grenade-Proof Greenhouse." *New York Times*, November 21, 2000.

Cohen, Stanley N. "The Manipulation of Genes." *Scientific American* (1975): 24–33.

Datta, S. K., et al. "Genetically Engineered Fertile Indica-Rice Recovered from Protoplasts." *Nature Biotechnology* 8 (1990): 736–40.

Fedoroff, Nina V., and Nancy Marie Brown. *Mendel in the Kitchen: A Scientist's View of Genetically Modified Foods.* Washington, DC: Joseph Henry Press, 2004.

Finer, John J., et al. "Development of the Particle Inflow Gun for DNA Delivery to Plant Cells." *Plant Cell Reports* (1992): 323–28.

International Rice Research Institute. *Biotechnology in International Agricultural Research.* Manila: IRRI, 1985.

International Service for the Acquisition of Agri-Biotech Applications (ISAAA). "Pocket K No. 13: Conventional Plant Breeding." Manila: Author, 2006. https://www.isaaa.org/resources/publications/pocketk/document/Doc-Pocket%20K13.pdf.

———. "Pocket K No. 37: Biotech Rice." Manila: Author, 2010. http://www.isaaa.org/resources/publications/pocketk/37/default.asp.

Khush, G. S., and G. H. Toenniessen, eds. *Rice Biotechnology.* Oxon: CAB International, 1991.

Margulis, Lynn, and Dorion Sagan. *What Is Life?* Berkeley: University of California Press, 2000.

National Public Radio. "In a Grain of Golden Rice, a World of Controversy over GMO Foods." March 7, 2013. https://www.npr.org/sections/thesalt/2013/03/07/173611461/in-a-grain-of-golden-rice-a-world-of-controversy-over-gmo-foods.

Potrykus, Ingo. "From the Concept of Totipotency to Biofortified Cereals." *Annual Review of Plant Biology* 66 (2015): 1–22. https://doi.org/10.1146/annurev-arplant-043014-114734.

———. "Golden Rice and Beyond." *Plant Physiology* 125 (2001): 1157–61.

———. "The Golden Rice 'Tale.'" *In Vitro Cellular & Developmental Biology—Plant* 37 (2001): 93–100. www.goldenrice.org/PDFs/The_GR_Tale.pdf.

Ricepedia. "Cultivated rice species." n.d. http://ricepedia.org/rice-as-a-plant/rice-species/cultivated-rice-species.

Rockefeller Foundation. "The Potential for Carotenoid Biosynthesis in Rice Endosperm." Summary report (unpublished), 1993.

Sanford, John C., et al. "Delivery of Substances into Cells and Tissues Using a Particle Bombardment Process." *Particulate Science and Technology* 5 (1987): 27–37.

CHAPTER 3 **GR 0.5 AND BEYOND**

Al-Babili, Salim, and Peter Beyer. "Golden Rice—Five Years on the Road—Five Years to Go?" *Trends in Plant Science* 10 (2005): 565–73. https://doi.org/10.1016/j.tplants.2005.10.006.

Dubock, Adrian. "The Present Status of Golden Rice." *Journal of Huazhhong Agricultural University* 33 (2014): 69–84. http://www.goldenrice.org/PDFs/Dubock-The_present_status_of_Golden_Rice-2014.pdf.

Greenpeace. "Genetically Engineered 'Golden Rice' Is Fools Gold." Press release, February 9, 2001. http://www.greenpeace.org/seasia/ph/press/releases/genetically-engineered-golden-2/.

Gura, T. "New Genes Boost Rice Nutrients." *Science* 13 (August 1999): 994–95.

Haerlin, Benedikt. Letter to the editor. *London Independent*, February 17, 2001, 2.

Hoa, T. T. C., et al. "Golden Indica and Japonica Rice Lines Amenable to Deregulation." *Plant Physiology* 133 (September 2003): 161–69. https://doi.org/10.1104/pp.103.023457.

Kryder, R. David, et al. "The Intellectual and Property Components of Pro-Vitamin A Rice (*Golden*Rice™): A Preliminary Freedom-to-Operate Review." *ISAAA Briefs*, no. 20 (2000). Ithaca, NY: ISAAA.

Martineau, Belinda. *First Fruit: The Creation of the Flavr Savr™ Tomato and the Birth of Genetically Engineered Food*. New York: McGraw-Hill, 2001.

Nash, J. Madeline. "This Rice Could Save a Million Kids a Year." *Time*, July 31, 2000. http://content.time.com/time/magazine/article/0,9171,997586,00.html.

Pollan, Michael. "The Way We Live Now: The Great Yellow Hype." *New York Times Magazine*, March 4, 2001.

Sheehy, Raymon E., et al. "Reduction of Polygalacturonase Activity in Tomato Fruit by Antisense RNA." *Proceedings of the National Academy of Sciences* 85 (1988): 8805–9.

Toenniessen, Gary. "How Golden Rice Got Its Name." YouTube video. Posted by International Rice Research Institute, September 30, 2009. https://www.youtube.com/watch?v=QzuV4HM-gJI.

Ye, X. D., et al. "Engineering the Provitamin A (Beta-Carotene) Biosynthetic Pathway into (Carotenoid-Free) Rice Endosperm." *Science* 28 (2000): 303–5.

CHAPTER 4 THE PROTOCOL

Applegate, John S. "The Prometheus Principle: Using the Precautionary Principle to Harmonize the Regulation of Genetically Modified Organisms." *Indiana Journal of Global Legal Studies* 9 (2001): 207–63. http://www.repository.law.indiana.edu/ijgls/vol9/iss1/11.

Berman, Steve W. "Thalidomide in America." *Seattlepi*, November 2, 2011. https://blog.seattlepi.com/steveberman/2011/11/02/thalidomide-in-america/.

Cartagena Protocol on Biosafety to the Convention on Biological Diversity. 2003. http://bch.cbd.int/protocol/text/.

Celgene Corporation. "Thalomid (thalidomide)." n.d. https://www.thalomid.com/.

Darnton, John. "Britain Ties Deadly Brain Disease to Cow Ailment." *New York Times*, March 21, 1996, A0001. https://www.nytimes.com/1996/03/21/world/brit ain-ties-deadly-brain-disease-to-cow-ailment.html.

Eisinger, François. "Precautionary Principle: A Self-Defeating Concept?" *Science* (2000).

European Commission. "Directive 2001/18/EC of the European Parliament and of the Council of 12 March 2001 on the Deliberate Release into the Environment of Genetically Modified Organisms and Repealing Council Directive 90/220/EEC—Commission Declaration." 2001. https://eur-lex.europa.eu/legal-content /en/TXT/?uri=CELEX:32001L0018.

Food and Drug Administration. "Preventable Adverse Drug Reactions: A Focus on Drug Interactions." n.d. https://www.fda.gov/drugs/developmentapproval process/developmentresources/druginteractionslabeling/ucm110632.htm.

Gurwitz, J. H., et al. "Incidence and Preventability of Adverse Drug Events in Nursing Homes." *American Journal of Medicine* 109, no. 2 (2000): 87–94.

Holm, Søren, and John Harris. "Precautionary Principle Stifles Discovery." *Nature* 400 (July 19, 1999): 398. https://doi.org/10.1038/22626.

Kelsey, Frances. "Autobiographical Reflections." 2014. https://www.fda.gov/down loads/AboutFDA/WhatWeDo/History/OralHistories/SelectedOralHistory Transcripts/UCM406132.pdf.

Klein, D. B., and A. Tabarrok. "Is the FDA Safe and Effective?" 2002. http://www .fdareview.org/05_harm.php.

Lazarou, Jason, et al. "Incidence of Adverse Drug Reactions in Hospitalized Patients." *Journal of the American Medical Association* 279 (1998): 1200–5.

National Cancer Institute. "Thalidomide." 2018. https://www.cancer.gov/about-can cer/treatment/drugs/thalidomide.

Parke, Emily C., and Mark A. Bedau. "The Precautionary Principle and Its Critics." In *The Ethics of Protocells*, edited by Mark A. Bedau and Emily C. Parke, 69–88. Cambridge, MA: MIT Press, 2009.

Redenbaugh, Keith, and Alan McHughen. "Regulatory Challenges Reduce Oppor- tunities for Horticultural Biotechnology." *California Agriculture* 58 (2004): 106–15. https://doi.org/10.3733/ca.v058n02p106.

Rio Declaration on Environment and Development. 1992. https://web.archive.org /web/20030402153036/http://habitat.igc.org/agenda21/rio-dec.htm.

Rouhi, Maureen. "Thalidomide." *Chemical & Engineering News* 83 (June 20, 2005). https://pubs.acs.org/cen/coverstory/83/8325/8325thalidomide.html.

Sunstein, Cass R. "Beyond the Precautionary Principle." *University of Pennsylvania Law Review* 151 (2003): 1003–58. https://scholarship.law.upenn.edu/penn_law _review/vol151/iss3/10.

Taleb, Nassim Nicholas, et al. "The Precautionary Principle (with Application to the Genetic Modification of Organisms)." Extreme Risk Initiative—NYU

School of Engineering Working Paper Series, September 4, 2014. https://arxiv
.org/abs/1410.5787.

Tantibanchachai, Chanapa. "US Regulatory Response to Thalidomide (1950–2000)."
Embryo Project Encyclopedia (January 4, 2014). http://embryo.asu.edu/handle
/10776/7733.

Vagani, Mauro, and Alessandro Olper. "Patterns and Determinants of GMO Regulations." *AgBioForum* 18 (2015): 44–54. http://www.agbioforum.org/v18n1/v18n1
a06-vigani.htm.

CHAPTER 5 **WHAT IS A GMO?**

Arber, Werner. "Genetic Engineering Compared to Natural Genetic Variations."
New Biotechnology 27 (November 2010): 517–21. https://doi.org/10.1016/j.nbt
.2010.05.007.

Berg, Paul, and Maxine Singer. *Dealing with Genes: The Language of Heredity.* Mill
Valley, CA: University Science Books, 1992.

Broad, William J. "Useful Mutants, Bred with Radiation." *New York Times*, August 28,
2007. https://www.nytimes.com/2007/08/28/science/28crop.html.

European Commission. "Directive 2001/18/EC of the European Parliament and of
the Council of 12 March 2001 on the Deliberate Release into the Environment
of Genetically Modified Organisms and Repealing Council Directive 90/220/
EEC—Commission Declaration." 2001. https://eur-lex.europa.eu/legal-content
/en/TXT/?uri=CELEX:32001L0018.

Feltman, Rachel. "Why This Genetically Modified Mushroom Gets to Skip USDA
Oversight." *Washington Post*, April 18, 2016.

Firko, Michael J. "Re: Request for Confirmation That Transgene Free, CRISPR-
Edited Mushroom Is Not a Regulated Article." USDA, April 13, 2016. https://
www.aphis.usda.gov/biotechnology/downloads/reg_loi/15-321-01_air_response
_signed.pdf.

Frank, Margaret H., and Daniel H. Chitwood. "Plant Chimeras: The Good, the
Bad, and the Bizzaria." *Developmental Biology* 419 (November 2016): 41–53.
https://doi.org/10.1016/j.ydbio.2016.07.003.

Goldschmidt, Eliezer E. "Plant Grafting: New Mechanisms, Evolutionary Implications." *Frontiers in Plant Science*, December 17, 2014. https://doi.org/10.3389
/fpls.2014.00727.

International Atomic Energy Agency. Joint FAO/IAEA Mutant Variety Database.
http://mvd.iaea.org.

Labrina. "Citrus aurantium 'bizzarria' frutto acerbo." *Wikimedia Commons.*
November 2, 2011. By Labrina—own work, CC BY-SA 3.0, https://commons
.wikimedia.org/w/index.php?curid=19015504.

Lewontin, Richard C. "Genes in the Food!" *New York Review of Books*, June 21, 2001. https://www.nybooks.com/articles/2001/06/21/genes-in-the-food/.

Melnyk, Charles W., and Elliot M. Meyerowitz. "Plant Grafting." *Current Biology* 25 (March 2, 2015): R183–R188. https://doi.org/10.1016/j.cub.2015.01.029.

Morton, J. "Grapefruit." In *Fruits of Warm Climates*, edited by Julia F. Morton, 152–58. Miami, FL: Echo, 1987. https://www.hort.purdue.edu/newcrop/morton /grapefruit.html.

Muller, H. J. "Artificial Transmutation of the Gene." *Science* 66 (July 22, 1927): 84–87. https://doi.org/10.1126/science.66.1699.84.

Onion, Amanda. "Fluorescent Potato Could Reduce Water Use." ABC News. https://abcnews.go.com/Technology/CuttingEdge/story?id=99368&page=1.

Ragionieri (Dr.). "Origin of the Florentine Bizzarria." *Journal of Heredity* (1937): 527–28. https://doi.org/10.1093/oxfordjournals.jhered.a102788.

Regis, Ed. "The Twilight Zone." In *What Is Life?*, by Ed Regis, 124–39. New York: Farrar, Straus and Giroux, 2008.

Rhoades, M. M. *Lewis John Stadler: A Biographical Memoir*. Washington, DC: National Academy of Sciences, 1957. http://www.nasonline.org/publications /biographical-memoirs/memoir-pdfs/stadler-lewis.pdf.

Rutgers, the State University of New Jersey. "National Survey Shows Americans Are in the Dark Regarding Genetically Modified Foods." *ScienceDaily*, February 1, 2005. https://www.sciencedaily.com/releases/2005/01/050131224504.htm.

Sikora, Per, et al. "Mutagenesis as a Tool in Plant Genetics, Functional Genomics, and Breeding." *International Journal of Plant Genomics* (2011). https://doi.org /10.1155/2011/314829.

Stegemann, Sandra, and Ralph Bock. "Exchange of Genetic Material between Cells in Plant Tissue Grafts." *Science* 324 (May 1, 2009): 649–51. https://doi.org/10.1126 /science.1170397.

Thompson and Morgan. "TomTato." n.d. https://www.thompson-morgan.com/p /tomtatoreg-ketchup-n-friestrade-ketchup-and-chips/t47176TM.

Zoler, Mitchel L. "At the Nation's Table: McAllen, Tex." *New York Times*, October 19, 1988. https://www.nytimes.com/1988/10/19/garden/at-the-nation-s-table-mc allen-tex.html.

CHAPTER 6 **SAFE TO EAT?**

Carson, Rachel. *Silent Spring*. 1962. New York: Houghton Mifflin, 2002.

Cellini, F., et al. "Unintended Effects and Their Detection in Genetically Modified Crops." *Food and Chemical Toxicology* 42 (July 2004): 1089–125. https://doi.org /10.1016/j.fct.2004.02.003.

Chassy, Bruce M. "Food Safety Risks and Consumer Health." *New Biotechnology* 27 (November 2010): 534–44. https://doi.org/10.1016/j.nbt.2010.05.018.

European Network of Scientists for Social and Environmental Responsibility. "No Scientific Consensus on GMO Safety." ENSSER Statement 21, October 2013.

Fedoroff, Nina V., and Nancy Marie Brown. *Mendel in the Kitchen: A Scientist's View of Genetically Modified Foods*. Washington, DC: Joseph Henry Press, 2004.

Food and Agriculture Organization. Codex Alimentarius. http://www.fao.org/fao-who-codexalimentarius/en/.

Jaffe, Ali. "How 12 EpiPens Saved My Life." *New York Times*, August 30, 2016. https://www.nytimes.com/2016/08/30/well/live/how-12-epipens-saved-my-life.html.

Karni, Moshe. "Thermal Degradation of DNA." *DNA and Cell Biology* 32 (2013): 1–4. https://doi.org/10.1089/dna.2013.2056.

Konnikova, Maria. "The Psychology of Distrusting G.M.O.s." *New Yorker*, August 8, 2013.

Kuntz, Marcel. "Destruction of Public and Governmental Experiments of GMO in Europe." *GM Crops & Food* 4 (October–December 2012): 258–64. https://doi.org/10.4161/gmcr.21231.

Latham, Jonathan R., et al. "The Mutational Consequences of Plant Transformation." *Journal of Biomedicine and Biotechnology* (2006): 1–7. https://doi.org/10.1155/JBB/200625376.

Lewis, Paul. Letter to the editor. *New York Times*, June 16, 1992, A00024. https://www.nytimes.com/1992/06/16/opinion/l-mutant-foods-create-risks-we-can-t-yet-guess-since-mary-shelley-332792.html.

Liu, Y. Z., et al. "Digestion of Nucleic Acids Starts in the Stomach." *Scientific Reports* 5 (2015). https://doi.org/10.1038/srep11936.

Losey, John E., et al. "Transgenic Pollen Harms Monarch Larvae." *Nature* 399 (May 20, 1999): 214. https://www.nature.com/articles/20338.

McHughen, Alan. *Pandora's Picnic Basket: The Potential and Hazards of Genetically Modified Foods*. New York: Oxford University Press, 2000.

National Academies of Sciences, Engineering, and Medicine. *Genetically Engineered Crops: Experiences and Prospects*. Washington, DC: National Academies of Sciences, Engineering, and Medicine, 2016.

Netherwood, T., et al. "Assessing the Survival of Transgenic Plant DNA in the Human Gastrointestinal Tract." *Nature Biotechnology* (2004): 204–9. https://doi.org/10.1038/nbt934.

Norero, Daniel. "More Than 280 Scientific and Technical Institutions Support the Safety of GM Crops." *Si Quiero Transgenicos*. Last updated June 19, 2017. http://www.siquierotransgenicos.cl/2015/06/13/more-than-240-organizations-and-scientific-institutions-support-the-safety-of-gm-crops.

Parsons, Sarah. "Study: GMO Crops Are Killing Butterflies." *Grist*, March 19, 2012. https://grist.org/animals/study-gmo-crops-are-killing-butterflies/.

Philpott, Tom. "Researchers: GM Crops Are Killing Monarch Butterflies After All." *Mother Jones*, March 21, 2012. https://www.motherjones.com/food/2012/03/researchers-gm-crops-are-killing-monarch-butterflies-after-all/.

Pleasants, J. M., and Oberhauser, K. S. "Milkweed Loss in Agricultural Fields because of Herbicide Use: Effect on the Monarch Butterfly Population." *Insect Conservation and Diversity* 6 (2013): 135–44. https://doi.org/10.1111/j.1752-4598.2012.00196.x.

Quaim, Matin. "Benefits of Genetically Modified Crops for the Poor: Household Income, Nutrition, and Health." *New Biotechnology* 27 (2010): 552–57. https://doi.org/10.1016/j.nbt.2010.07.009.

Sears, Mark K., et al. "Impact of *Bt* Corn Pollen on Monarch Butterfly Populations: A Risk Assessment." *Proceedings of the National Academy of Sciences* 98 (October 9, 2001): 11937–42. https://doi.org/10.1073/pnas.211329998.

Society of Toxicology. "The Safety of Genetically Modified Foods Produced through Biotechnology." *Toxicological Sciences* 71 (2003): 2–8. https://doi.org/10.1093/toxsci/71.1.2.

War, A. R., et al. "Mechanisms of Plant Defense against Insect Herbivores." *Plant Signaling & Behavior* (2012): 1306–20.

Weasel, Lisa. *Food Fray: Inside the Controversy over Genetically Modified Food.* New York: AMACOM, 2009.

Yoon, Carol Kaesuk. "Altered Corn May Imperil Butterfly, Researchers Say." *New York Times*, May 20, 1999, A1.

CHAPTER 7 GOLDEN RICE 2

Chassy, Bruce. "Food Safety of Transgenic Rice." In *Rice Improvement in the Genomics Era*, edited by S. K. Datta, 417–55. London: Taylor and Francis, 2009.

———. "Golden Rice 2." *Comprehensive Reviews in Food Science and Food Safety* 7 (2008): 92–98. https://onlinelibrary.wiley.com/doi/abs/10.1111/j.1541-4337.2007.00029_7.x.

Datta, Swampan K., et al. "Golden Rice: Introgression, Breeding, and Field Evaluation." *Euphytica* 154 (2007): 271–78. https://doi.org/10.1007/s10681-006-9311-4.

Dubock, Adrian. "The Present Status of Golden Rice." *Journal of Huazhhong Agricultural University* 33 (2014): 69–84. http://www.goldenrice.org/PDFs/Dubock-The_present_status_of_Golden_Rice-2014.pdf.

Food and Drug Administration. "Statement of Policy: Foods Derived from New Plant Varieties." *Federal Register* 57, no. 104 (May 29, 1992): 22984. https://

www.fda.gov/Food/GuidanceRegulation/GuidanceDocumentsRegulatoryInfor
mation/Biotechnology/ucm096095.htm.

Golden Rice Project. http://www.goldenrice.org/.

Goodman, G. E. "The Beta-Carotene and Retinol Efficacy Trial." *Journal of the National Cancer Institute* 96 (2004): 1743–50. https://doi.org/10.1093/jnci/djh320.

Goodman, Richard E., and John Wise. "Bioinformatic Analysis of Proteins in
Golden Rice 2 to Assess Potential Allergenic Cross-Reactivity." University of
Nebraska, Food Allergy Research and Resource Program, study no. BIO-02-
2006. 2006. http://www.allergenonline.org/Golden%20Rice%202%20Bioinfor
matics%20FARRP%202006.pdf.

Greenpeace. *Golden Illusion: The Broken Promises of "Golden Rice."* Amsterdam:
Greenpeace International, 2013.

———. "Renewed Golden Rice Hype Is Propaganda for Genetic Engineering Industry." Greenpeace Philippines, feature story, March 21, 2005. http://www.green
peace.org/seasia/ph/News/news-stories/renewed-golden-rice-hype-is-pr/.

Hoa, T. T. C., et al. "Golden Indica and Japonica Rice Lines Amenable to Deregulation." *Plant Physiology* 133, no. 1 (September 2003): 161–69. https://doi.org
/10.1104/pp.103.023457.

Linscombe, Steven. "Golden Rice Could Help Malnutrition." LSU AgCenter, news
release, October 13, 2004.

———. "Nutritional Benefits of Golden Rice May Not Be Realized for Several More
Years." LSU AgCenter, news release, December 15, 2005.

Paine, Jacqueline A., et al. "Improving the Nutritional Value of Golden Rice
through Increased Pro-vitamin A Content." *Nature Biotechnology* 23 (2005):
482–87. https://doi.org/10.1038/nbt1082.

Potrykus, Ingo. "From the Concept of Totipotency to Biofortified Cereals." *Annual
Review of Plant Biology* 66 (2015): 1–22. https://doi.org/10.1146/annurev-arplant
-043014-114734.

———. "Golden Rice and Beyond." *Plant Physiology* 125 (2001): 1157–61.

———. "The Golden Rice 'Tale.'" *In Vitro Cellular & Developmental Biology—Plant*
37 (2001): 93–100. http://www.goldenrice.org/PDFs/The_GR_Tale.pdf.

Rice Research Station Newsletter. "Golden Rice Research." LSU AgCenter 1,
November 1, 2004.

Virk, Parminder. "Golden Rice as New Varieties." In *Golden Rice*, 17–21. Stockholm: Royal Swedish Academy of Agriculture and Forestry, 2009.

CHAPTER 8 **BETTER THAN SPINACH**

Allow Golden Rice Now! Society. http://allowgoldenricenow.org.

Brown, David, and David Sapsted. "GM Battle Fears as Melchett Is Cleared." *Tele-
graph*, September 21, 2000.

Dubock, Adrian. "'The Politics of Golden Rice." *GM Crops & Food* 5 (2014): 210–22. https://doi.org/10.4161/21645698.2014.967570.

Enserink, Martin. "Golden Rice Not So Golden for Tufts." *Science*, September 18, 2013.

———. "Researcher Sues to Block Retraction of Golden Rice Paper." *Science*, July 17, 2014.

Greenpeace East Asia. "GE Rice Paper Retracted from the *American Journal of Clinical Nutrition*—Greenpeace Comment." Press release, August 11, 2015.

———. "24 Children Used as Guinea Pigs in Genetically Engineered 'Golden Rice' Trial." Greenpeace East Asia (blog), 2012. http://www.greenpeace.org/eastasia/news/blog/24-children-used-as-guinea-pigs-in-geneticall/blog/41956/.

Haerlin, Benedikt. Letter to the editor. *London Independent*, February 17, 2001, 2.

Hvistendahl, Mara, and Martin Enserink. "Chinese Researchers Punished for Role in GM Rice Study." *Science*, December 12, 2010. https://www.sciencemag.org/news/2012/12/chinese-researchers-punished-role-gm-rice-study.

Moore, Patrick. *Confessions of a Greenpeace Dropout: The Making of a Sensible Environmentalist*. Vancouver, BC: Beatty Street, 2010.

———. "Why I Left Greenpeace." *Wall Street Journal*, April 22, 2008. https://www.wsj.com/articles/SB120882720657033391.

Ribaya-Mercado, J. D., et al. "Carotene-Rich Plant Foods Ingested with Minimal Dietary Fat Enhance the Total-Body Vitamin A Pool Size in Filipino School-children as Assessed by Stable-Isotope-Dilution Methodology." *American Journal of Clinical Nutrition* 85, no. 4 (April 2007): 1041–49. https://doi.org/10.1093/ajcn/85.4.1041.

Stone, G. D., and D. Glover. "Disembedding Grain: Golden Rice, the Green Revolution, and Heirloom Seeds in the Philippines." *Agriculture and Human Values* 34 (2017): 87–102. https://doi.org/10.1007/s10460-016-9696-1.

Tang, Guangwen, et al. "ß-Carotene in Golden Rice Is as Good as ß-Carotene in Oil at Providing Vitamin A to Children." *American Journal of Clinical Nutrition* 96 (2012): 658–64. https://doi.org/10.3945/ajcn.111.030775.

———. "Golden Rice Is an Effective Source of Vitamin A." *American Journal of Clinical Nutrition* 89 (2009): 1776–83. https://doi.org/10.3945/ajcn.2008.27119.

———. "Spinach or Carrots Can Supply Significant Amounts of Vitamin A as Assessed by Feeding with Intrinsically Deuterated Vegetables." *American Journal of Clinical Nutrition* 82 (2005): 821–28.

Virk, Parminder. "Golden Rice as New Varieties." In *Golden Rice*, 17–21. Stockholm: Royal Swedish Academy of Agriculture and Forestry, 2009.

Zeigler, Robert S. "Biofortification: Vitamin A Deficiency and the Case for Golden Rice." In *Plant Biotechnology: Experience and Future Prospects*, edited by A. Ricroch et al., 245–62. Switzerland: Springer International, 2014. https://doi.org/10.1007/978-3-319-06892-3_19.

CHAPTER 9 **THE MISTAKE**

Ahmad, Reaz. "Bangladeshi Scientists Ready for Trial of World's First 'Golden Rice.'" *Daily Star*, October 8, 2015.

——. "Vitamin A Rice Now a Reality." *Daily Star*, October 28, 2016.

Alfonso, Antonio A. "The Deployment and Validation of High Beta-Carotene Rice Varieties in the Philippines and Bangladesh to Combat Vitamin A Deficiency." In *Philippine Rice R&D Highlights 2012*. Muñoz, Nueva Ecija, Philippines: Phil-Rice, 2012.

Asis, Marlo. "Golden Rice Moving forward in Philippines." Ithaca, NY: Cornell Alliance for Science, August 7, 2017. https://allianceforscience.cornell.edu/blog/2017/08/golden-rice-moving-forward-in-philippines/.

Batista, Rita, et al. "Microarray Analyses Reveal That Plant Mutagenesis May Induce More Transcriptomic Changes Than Transgene Insertion." *Proceedings of the National Academy of Sciences* 105 (March 2008): 3640–45. https://doi.org/10.1073/pnas.0707881105.

Baudo, M. M., et al. "Transgenesis Has Less Impact on the Transcriptome of Wheat Grain Than Conventional Breeding." *Plant Biotechnology Journal* 4 (2006): 369–80. https://doi.org/10.1111/j.1467-7652.2006.00193.x.

Biswas, Partha S., et al. *Recent Advances in Breeding Golden Rice in Bangladesh*. International Rice Research Institute. Los Baños, Philippines: IRRI, 2016. http://ilsirf.org/wp-content/uploads/sites/5/2016/09/Biswas-P.pdf.

Bollenedi, Haritha, et al. "Molecular and Functional Characterization of GR2-R1 Event Based Backcross Derived Lines of Golden Rice in the Genetic Background of a Mega Rice Variety Swarna." *PLoS ONE* 12 (January 9, 2017): e0169600. https://doi.org/10.1371/journal.pone.0169600.

Deb, Uttam. *Returns to Golden Rice Research in Bangladesh: An Ex-ante Analysis*. Copenhagen: Bangladesh Priorities, Copenhagen Consensus Center, 2016. https://www.copenhagenconsensus.com/sites/default/files/deb_golden_rice.pdf.

Dubock, Adrian. "The Present Status of Golden Rice." *Journal of Huazhhong Agricultural University* 33 (2014): 69–84. http://www.goldenrice.org/PDFs/Dubock-The_present_status_of_Golden_Rice-2014.pdf.

Harmon, Amy. "Golden Rice: Lifesaver?" *New York Times*, August 24, 2013.

International Service for the Acquisition of Agri-Biotech Applications. "ISAAA Brief 41-2009: Executive Summary." 2009. http://www.isaaa.org/resources/publications/briefs/41/executivesummary/default.asp.

Lynas, Mark. "The True Story about Who Destroyed a Genetically Modified Rice Crop." *Slate*, August 26, 2013.

PhilRice and IRRI. *Compilation of Study Reports: Studies Submitted in Support of the Food Safety Assessment of Provitamin A Biofortied GR2E Rice*. Muñoz, Nueva Ecija, Philippines: PhilRice; Los Baños, Philippines: IRRI, 2017.

———. *Provitamin A Biofortified Rice Event GR2E*. Muñoz, Nueva Ecija, Philippines: PhilRice; Los Baños, Philippines: IRRI, 2017.

Slezak, Michael. "Militant Filipino Farmers Destroy Golden Rice Crop." *New Scientist*, August 9, 2013.

Then, Christoph. "'Golden Rice': Unexpected Genomic Effects." TestBiotech.org, February 15, 2017.

Wilson, Allison. "Goodbye to Golden Rice? GM Trait Leads to Drastic Yield Loss and 'Metabolic Meltdown.'" *Independent Science News*, October 25, 2017.

CHAPTER 10 THE "CRIME AGAINST HUMANITY"

Achenbach, Joel. "107 Nobel Laureates Sign Letter Blasting Greenpeace over GMOs." *Washington Post*, June 30, 2016.

Allow Golden Rice Now! Society. http://allowgoldenricenow.org.

Chokshi, Niraj. "Stop Bashing G.M.O. Foods, More Than 100 Nobel Laureates Say." *New York Times*, June 30, 2016.

GMWatch. "Philippines Supreme Court Reverses Bt Eggplant Ruling." July 29, 2016. https://www.gmwatch.org/en/news/latest-news/17125-philippines-supreme-court-reverses-bt-eggplant-ruling.

Greenpeace Manila. "Nobel Laureates Sign Letter on Greenpeace 'Golden' Rice Position—Statement." June 30, 2016. https://www.greenpeace.org/international/press-release/6866/nobel-laureates-sign-letter-on-greenpeace-golden-rice-position-statement/.

Hutchinson, Brian. "Ex-Greenpeace President Says Group's Opposition to Genetically-Modified Golden Rice Costing Thousands of Lives." *National Post*, October 11, 2013.

Lee, Hyejin, and Sheldon Krimsky. "The Arrested Development of Golden Rice: The Scientific and Social Challenges of a Transgenic Biofortified Crop." *International Journal of Social Science Studies* 4 (November 2016): 51–64. https://doi.org/10.11114.ijsss.v4i11.1918.

Moore, Patrick. "The Crime against Humanity." n.d. http://allowgoldenricenow.org/the-crime-against-humanity/.

Paarlberg, Robert. "A Dubious Success: The NGO Campaign against GMOs." *GM Crops & Food* (2014): 223–28.

Pelegrina, D. "Greenpeace Wins Landmark GE Eggplant Court Case." Greenpeace (blog), May 14, 2012.

Phillips McDougall. *The Cost and Time Involved in the Discovery, Development and Authorization of a New Plant Biotechnology Derived Trait*. Midlothian, UK:

Author, September 2011. https://croplife.org/wp-content/uploads/pdf_files/Get ting-a-Biotech-Crop-to-Market-Phillips-McDougall-Study.pdf.

Potrykus, Ingo. "Golden Rice and the Greenpeace Dilemma." *AgBioWorld*, February 15, 2001.

———. "Lessons from the 'Humanitarian Golden Rice' Project: Regulation Prevents Development of Public Good Genetically Engineered Crop Products." *New Biotechnology* 27 (2010): 466–72. https://doi.org/10.1016/j.nbt.2010.07.012.

Stein, Alexander J., et al. "Genetic Engineering for the Poor: Golden Rice and Public Health in India." *World Development* 36 (2008): 144–58. https://doi.org /10.1016/j.worlddev.2007.02.013.

STOP Golden Rice Alliance. "Golden Rice Is Unnecessary and Dangerous." March 10, 2015. https://www.grain.org/article/entries/5177-golden-rice-is-unnecessary -and-dangerous.

US Department of Agriculture. *Report of LibertyLink Rice Incidents*, 2007. https:// www.usda.gov/sites/default/files/documents/LLP%20Incidents%202.docx.

Wessler, Justus, and David Zilberman. "The Economic Power of the Golden Rice Opposition." *Environment and Development Economics* 19 (2013): 724–42. https://doi.org/10.1017/S1355770X1300065X.

Zimmerman, Z., and M. Qaim. "Potential Health Benefits of Golden Rice: A Philippine Case Study." *Food Policy* 29 (2004): 147–68. https://doi.org/10.1016/j .foodpol.2004.03.001.

CHAPTER 11 **THE APPROVALS**

Food Standards Australia New Zealand. *Approval Report—Application A1138: Food Derived from Provitamin A Rice Line GR2E*. December 20, 2017. http://www .foodstandards.gov.au/code/applications/Pages/A1138GMriceGR2E.aspx.

Genetic Literacy Project. http://geneticliteracyproject.org.

Government of Canada, Health Canada. "Provitamin A Biofortified Rice Event GR2E (Golden Rice)." March 16, 2018. https://www.canada.ca/en/health-can ada/services/food-nutrition/genetically-modified-foods-other-novel-foods /approved-products/golden-rice-gr2e.html.

Keefe, Dennis S., to Donald J. MacKenzie, May 24, 2018. "RE: Biotechnology Notification File No. BNF 000158." US Food and Drug Administration. https:// www.fda.gov/downloads/Food/IngredientsPackagingLabeling/GEPlants/Sub missions/ucm608797.pdf.

Lynas, Mark. "Anti-GMO Activists Convene to Target Golden Rice." Cornell Alliance for Science, April 4, 2018. https://allianceforscience.cornell.edu/blog/2018 /04/anti-gmo-activists-convene-target-golden-rice/.

MASIPAG. Press release. April 20, 2018. http://masipag.org/2018/04/.

McMahon, Carrie. "Biotechnology Notification File No. 000158: Note to the File." US Food and Drug Administration. May 8, 2018. https://www.fda.gov/down loads/Food/IngredientsPackagingLabeling/GEPlants/Submissions/ucm607450 .pdf.

PhilRice and IRRI. "Supporting Information of the Risk Analysis Report for a Genetically Modified Plant for Direct use as Food, Feed, or Processing." *Provitamin A Biofortified Rice Event GR2E*. Muñoz, Nueva Ecija, Philippines: PhilRice; Los Baños, Philippines: IRRI, 2017.

Stop Golden Rice! Network. "Civil Society Decries FSANZ Approval of Golden Rice." Grain.org, February 22, 2018. https://www.grain.org/article/entries/5897 -civil-society-decries-fsanz-approval-of-golden-rice.

Then, Christoph. "Data on 'Golden Rice' Not Sufficient to Show Health Safety and Indicate Low Benefits." TestBiotech.org, February 5, 2018. https://www.testbio tech.org/en/news/data-golden-rice-not-sufficient-show-health-safety-and-indi cate-low-benefits.

Wan, Lester. "GE-Free New Zealand Planning to Sue Authorities over Golden Rice Approval." FoodNavigator-asia.com, March 1, 2018. https://www.foodnavigator -asia.com/Article/2018/03/01/GE-Free-New-Zealand-planning-to-sue-authori ties-over-golden-rice-approval#.

Wilson, Allison. "Goodbye to Golden Rice? GM Trait Leads to Drastic Yield Loss and 'Metabolic Meltdown.'" *Independent Science News*, October 25, 2017. Comment, March 24, 2018.

Wilson, Allison, and Jonathan Latham. "GMO Golden Rice Offers No Nutritional Benefits Says FDA." *Independent Science News*, June 3, 2018.

EPILOGUE **THE PROACTIONARY PRINCIPLE**

GMWatch. "Uganda: Museveni Declines to Sign GMO Bill into Law." December 29, 2017. https://www.gmwatch.org/en/news/archive/18048-uganda-museveni-de clines-to-sign-gmo-bill-into-law.

More, Max. "The Perils of Precaution." Max More's Strategic Philosophy (blog), August 22–23, 2010. http://strategicphilosophy.blogspot.com/2010/08/.

———. "The Proactionary Principle." Max More's Strategic Philosophy (blog), March 28, 2008. http://strategicphilosophy.blogspot.com/2008/03/proaction ary-principle-march-2008.html.

———. "The Proactionary Principle." July 29, 2005. http://strategicphilosophy.blog spot.com/2010/08/.

Paul, Jean-Yves, et al. "Golden Bananas in the Field: Elevated Fruit Pro-vitamin A from the Expression of a Single Banana Transgene." *Plant Biotechnology Journal* 15 (October 13, 2016): 520–32. https://doi.org/10.1111/pbi.12650.

Potrykus, Ingo. "Regulation Must Be Revolutionized." *Nature* 466 (July 29, 2010): 561. https://doi.org/10.1038/466561a.
Queensland University of Technology. "Golden Bananas High in Pro-vitamin A Developed: Research Has Produced a Golden-Orange Fleshed Banana, Rich in Pro-vitamin A." *ScienceDaily*, July 7, 2017. www.sciencedaily.com/releases/2017 /07/170707095806.htm.

APPENDIX *L'AFFAIRE SCHUBERT*

Beachy, Robert, et al. "Different Perspectives on GM Food." *Nature Biotechnology* 20 (December 2002): 1195.
Schaub, Patrick, et al. "Nonenzymatic β-Carotene Degradation in Provitamin A–Biofortified Crop Plants." *Journal of Agricultural and Food Chemistry* 65 (2017): 6588–98. https://doi.org/10.1021/acs.jafc.7b01693.
Schubert, David. "A Different Perspective on GM Food." *Nature Biotechnology* 20 (October 2002): 969.
Schubert, David R. "The Problem with Nutritionally Enhanced Plants." *Journal of Medicinal Food* 11 (2008): 601–5. https://doi.org/10.1089/jmf.2008.0094.

INDEX